Hill Country Landowner's Guide

Number Forty-four:
Louise Lindsey Merrick Natural Environment Series

Hill Country Landowner's Guide

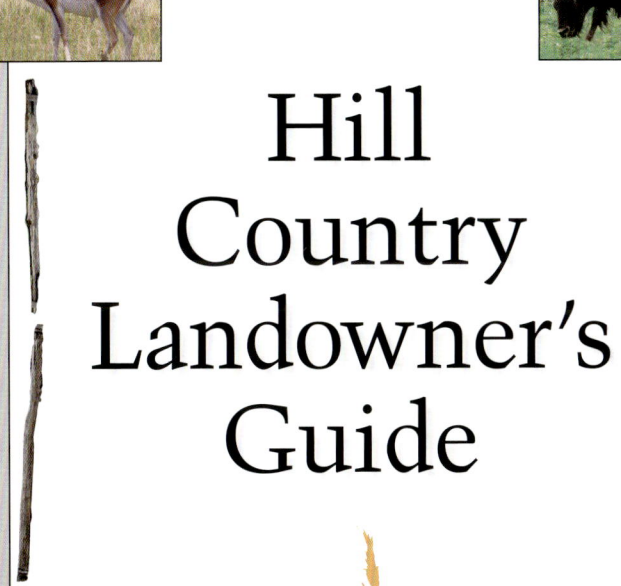

Jim Stanley

Texas A&M University Press

College Station

LIBRARY OF CONGRESS CATALOGING-IN-PUBLICATION DATA

Stanley, Jim, 1941—
Hill Country landowner's guide / Jim Stanley, — 1st ed.
p. cm. — (Louise Lindsey Merrick natural environment series ; no. 44)
Includes bibliographical references and index.
ISBN - 13: 978-1-60344-137-7 (flexbound with flaps : alk. paper)
ISBN - 10: 1-60344-137-9 (flexbound with flaps : alk. paper)
1. Land use, Rural Environmental aspects—Texas—Texas Hill Country—
Handbooks, manuals, etc. 2. Landscape protection—Texas—Texas Hill Country—
Handbooks, manuals, etc. 3. Natural resources management—Texas—
Texas Hill Country—Handbooks, manuals, etc. 4. Ecosystem management—
Texas—Texas Hill Country—Handbooks, manuals, etc. I. Title. II. Series:
Louise Lindsey Merrick natural environment series ; no. 44
HD211.T4S73 2009
333.7909764'3—dc22

2009007535

To Priscilla,

the love of my life and without whom this book

would never have been written.

And to the landowners

of the Texas Hill Country who have the ability

to maintain this part of the world in a biologically

healthy condition, if they so choose.

Contents

Preface

The Texas Hill Country has seen and continues to see a great influx of new landowners, moving here from all over the world. Many of these new landowners, instead of moving into the various towns and cities, have chosen to buy a piece of property in the rural areas. Most of these properties have been created by dividing larger, longtime ranches into smaller parcels. Many of these newcomers have little if any experience in managing a piece of ranchland, here or anywhere else. It is primarily for these new landowners that this book is intended.

This is a handbook for Texas Hill Country landowners. The farther away from the Hill Country one lives, the less relevant some of the discussion will be, so folks in far East Texas or the Panhandle won't get much specific useful information from this book. The book is also aimed at those Hill Country landowners who own, for want of better terms, ranchland or native pastures, whether or not they own any grazing animals. This book does not address any of the issues involved in managing orchards, vineyards, row crops, or introduced grass hay crops.

Having visited many landowners who fit the description above, as well as understanding the importance that all Hill Country land be well managed for the benefit of us all, I have seen a need for an easy-reference guide to help people with their land management decisions. That is my main purpose in writing this book.

There is always a compromise in writing any nonfiction book in terms of how much detail and breadth to include. I hope I have struck the right balance. I also decided early on that, although most of what is discussed is science based, the writing should be neither scientific jargon nor overly simplified, but rather a conversational middle road that gives enough explanation of the why and how without too much detail. Only you can decide if I succeeded.

About the Author

Most readers of any nonfiction book naturally wonder about the author's credentials. Does she know enough about the subject to be writing about it? Is he proselytizing an agenda or a point of view? To help readers judge for themselves, here is my biography:

I grew up in an oil camp outside Seminole, Texas, in Gaines County, between Lubbock and Midland. As a kid I worked on the cotton farm next to the camp and raised 4-H livestock. I went to Texas Tech and received a BS and an MS in chemistry and then earned a PhD from Indiana University. I spent seven years at Louisiana State University doing postdoctoral research and teaching undergraduate chemistry. I then went to work for Union Carbide Corporation in their research and development labs in New York and New Jersey and retired from a position as senior research scientist in 2000.

Since retiring to the Texas Hill Country in 2000, my wife Priscilla and I have immersed ourselves in all aspects of the Hill Country ecology, joining the Kerrville chapter of the Native Plant Society of Texas, Riverside Nature Center in Kerrville, and Cibolo Nature Center in Boerne. We were in the first class of the Hill Country chapter of the Texas Master Naturalists, and I was the second president of that organization and have now served three terms. The Texas Master Naturalists program is patterned after the Master Gardener program and is sponsored by the Texas Parks and Wildlife Department and the Texas AgriLife Extension Service. Becoming a certified Texas Master Naturalist requires an initial forty-plus hours of classroom instruction on various nature topics including geology; archeology; birds; mammals; deer; fish; reptiles; woodlands, trees, and shrubs; grasslands and grasses; riparian areas; ecosystems; and range management, followed by eight hours of advanced training and forty hours of volunteer work. An additional eight hours of advanced training and forty hours of volunteer work are required every year.

In the process becoming a master naturalist but also taking advantage of every possible opportunity to learn about all aspects of the Hill Country ecology, I have attended more than four hundred hours of classes, seminars, conferences, workshops, and field days, presented by recognized experts, many of whom are listed in chapter 2. Since most of these events have involved trips to various ranches or other properties, I have visited more than seventy-five such properties in the Hill Country. In addition, as part of my work with the Texas Master Naturalists, I developed a program whereby, after additional special training, volunteers visit landowners' properties, upon request, and help identify the plants growing there and discuss various land management issues in an effort to help landowners achieve their goals. We have visited more than 160 properties in Bandera, Gillespie, Kendall, and Kerr counties since the beginning of this Land Management Assistance Program.

Although one can learn a lot from reading books and articles, and I have done a lot of that, the vast majority of what I know about the Hill Country has come from the real experts, the biologists, range scientists, foresters, and other specialists working for such government agencies as the Texas AgriLife Extension, Texas Parks and Wildlife, Texas Forest Service, and USDA Natural Resources Conservation Service, along with many faculty members from various universities. I have found the vast majority of these people to be extremely knowledgeable, hard working, dedicated, passionate about their work, and always eager to share their knowledge with others.

JIM STANLEY
Kerrville, Texas

Hill Country Landowner's Guide

I

The Philosophy

In the end, we will conserve only what we love, we will love only what we understand and we will understand only what we have been taught. —Bada Diom

I PRESUME that if you have opened this book and have come to this page, we have certain things in common. It is probably safe to assume that you own a piece of property, probably in the Texas Hill Country, and that you have at least some questions or concerns about how to manage it. It is with those assumptions that this book is written.

We all agree, I believe, that if you bring children into this world you have a certain moral obligation to nurture and protect them for as long as necessary. Likewise, I would argue, if you adopt a puppy or a kitten you have a similar obligation. I would further suggest that a somewhat similar obligation pertains to buying a piece of land. People and puppies have finite lifetimes, but the land lives on forever, so our actions with respect to the land have a longer-lasting effect than how we raise our kids or pets.

Being a good steward of the land should be as much a require-ment of a good citizen as being a good parent. Regardless of what the laws and the books in the courthouse say, we don't really "own" the land, we are just the current tenants who are taking care of the place for a brief time before passing it on to the next generation. Common sense and common courtesy require that we leave the land in at least as good a condition as we found it.

The beauty of the Hill Country is just too special to lose, we must preserve it.

As Aldo Leopold, famed ecologist, conservationist, and environmentalist, wrote in *A Sand County Almanac* in 1949, "We abuse land because we consider it a commodity belonging to us. When we see land as a community to which we belong, we may begin to use it with love and respect. There is no other way for land to survive the impact of mechanized man. . . . That land is a community is a basic concept of ecology, but that land is to be loved and respected is an extension of ethics."

Professor Larry White, of Texas AgriLife Extension at Texas A&M, puts the idea somewhat differently:

> As a rancher/landowner you manage an entire ecosystem of interrelated factors and resources. Some factors you can control, others you "learn" to live within constraints or suffer the consequences. Your Land Ethics and Stewardship Goals determine what you select to do.
>
> Land Ethics is the moral philosophy, standards of conduct and moral judgment related to the land/natural resources/environment.
>
> Land Stewardship is assuming the responsibilities for the care and use of the land resource.

Texans have a strong attitude that "It's my land and I will do what I want to on it." As a consequence, we have few laws regulating what you can in fact do with your land. Zoning laws are pretty much nonexistent in rural areas, as are most any other legal restrictions. One could argue that this is good or bad, but in reality it is what it is.

About 95 percent of the land area in the state of Texas is privately owned. In spite of some great state parks and magnificent national parks, we have a smaller percentage of public land than most other states. The relevance of this fact is, simply, that if land is to be conserved and managed well, it will have to be done by private landowners. And given the lack of any rules or restrictions in this area, it will have to be done voluntarily. Thus, to use Professor White's terms, the land ethics and land stewardship of individual landowners are critical to keeping Texas looking like Texas.

In the coming chapters, I discuss in detail why we all have an interest in how everyone else is managing their land. But for now a quote from Lyndon Johnson, in 1947, before he ever went to the White House, sums it up best: "Saving the water and the soil must start where the first raindrop falls."

We proceed with the assumption that there is at least general agreement as to the ethical responsibilities attendant to owning a piece of property, and that we all want to "take care" of our land, to "manage it well," to "leave it in better shape than we found it." But well-meaning people with the best of intentions can do bad things if they don't know what is, in fact, good and bad land management practice. And before we can talk about the details, we have to have at least a broad general picture of what we think our land should look like, what the ideal Hill Country landscape should be. What is our goal?

When asked that question, some intelligent, well-meaning folks might answer that they want "nature to take its course." This sounds good; who could be against nature, after all? The problem is that this is not really a vision of what they want the land to look like, but rather a laissez-faire management style that accepts whatever results. In most areas of the Hill Country, what would result is a cedar brake.

Other folks might answer that they want the land to look like it did before Europeans began settling the area, that is, before about 1830. At first thought, this sounds like a great vision. But to do that, we would have to stop farming and let the land revert to grassland, take down all the fences, and bring back huge herds of bison. We would also have to bring back the black bear, the wolf, and larger numbers of mountain lions. We would have to let wildfires burn themselves out. Clearly, that isn't going to happen.

So what would describe an ideal Hill Country landscape? I think the best answer is to think about the land as a biologist would and invoke two of the most powerful concepts in biology and ecology: diversity and sustainability. *Diversity* has to do with variety, in terms of both numbers of plant and animal species and sizes and ages of the longer-lived species. *Sustainability* has to do with the ability of the ecosystem to continue long term in the current state. My dictionary defines *sustainable* in this context as, "a method of harvesting or using a resource so that the resource is not depleted or permanently damaged." An ecosystem is sustainable if there is a balance of the numbers of each species so that consumers only consume as much as the land can produce over the long term. No one species crowds out any others, no species becomes overpopulated, and none is eliminated.

You will notice that the above definitions do not list which species should or should not be present or any specific stocking rate or crop. That is because there may be many different collections of species that will work in different ecosystems throughout the Hill Country. No two pieces of property are exactly alike; no two will have exactly the same percentage of each species.

The definition of an ideal Hill Country landscape below is my attempt to put into words a description of my vision. Yours may be quite different. That's okay. In fact, that's good. And even if you can't put your vision into words, if you at least have a general mental picture of what you want your place to become, or continue to be, then you can be in a position to judge each possible management activity in terms of your vision. And that is what this book is designed to help you do.

My ideal Hill Country landscape would

- Consist of a high level of native vegetative diversity, of a quality and quantity that are sustainable long term and that captures rainwater and prevents soil erosion.
- Support natural populations of native animals by providing food, water, shelter, and cover without degrading the habitat.
- Contain a mixture of mid- and tall grasses interspersed with various forbs (broad-leaved herbaceous plants, wildflowers, and weeds) and many different species of trees and understory shrubs.

Finally, as the population of the Hill Country increases and the rainfall doesn't, we are going to be increasingly concerned about our water supply. In later chapters I discuss why land management affects water quality and quantity, but suffice it to say at this point that how we all manage our properties affects all the rest of us in terms of our water supply. Good land management tends to improve water quality and quantity; poor land management does the opposite. So we all have a stake in what our neighbor is doing on his land, and we all have an obligation to our neighbors to take care of our own land.

2

The Real Experts

MUCH of what I have learned about Hill Country ecology has come from the individuals listed in this chapter. One of the first things I learned when I began attending events featuring these experts as speakers is how easily approachable and helpful they are and how much they really want to share their knowledge. The other thing I learned early on was, somewhat surprisingly, how much near total agreement there was among these experts even though they worked for different agencies with different missions.

I have included their names in a chapter of this book rather than in the acknowledgments to make sure that everyone who reads this book sees them. Unfortunately, I am sure that I have left some people off the lists that should have been included. It is impossible to list all of the many informative presentations and interesting discussions I have had over the past several years, all of which contributed greatly to both my knowledge and my enthusiasm for the subject of Hill Country ecology.

Another reason for including their names here is to illustrate how many experts are available in the region to help landowners with land management issues. Contact information is provided in chapter 23.

Texas Parks and Wildlife Department
  Bill Armstrong (ret.), Kerr Wildlife Management Area
  Mitch Lockwood, Kerrville
  Joyce Moore, Harper
  Dale Prochaska, Kerr Wildlife Management Area
  Rufus Stephens, Boerne
  Max Traweek (ret.), Kerrville
Texas AgriLife Extension
  Billy Kniffen, Menard County
  Bob Lyons, Uvalde Research Station
  Barron Rector, Texas A&M
  Roy Walston, Kerr County
  Larry White, Texas A&M
  Brad Wilcox, Texas A&M
Texas Forest Service
  Mark Duff, Kerrville
  Robert Edmonson, Johnson City
  Susan Sander, Kerrville
USDA/Natural Resources Conservation Service
  Joe Franklin, Kerrville
  Lee Knox, formerly Kerrville
  Steve Nelle, San Angelo
Bamberger Ranch
  David and Margaret Bamberger
  Lew Hunnicutt
(There are many, many others.)

One final point. Many environmental groups have as their mission, not only a certain segment of nature or the environment, but also a political agenda to further their cause. I find no fault with most of these groups and even share many of their goals. But my purpose, that of the Texas Master Naturalists, and that of most of the agency experts in the lists above is to provide factual, scientific information based on the best consensus of experts in the field without any political, religious, or other agenda. Biology, ecol-

Bob Lyons of the Texas AgriLife Extension Service discusses land management with members of the Hill Country chapter of the Texas Master Naturalists, including the author (in the big hat).

ogy, land management, and all of the related fields are sciences, which means that the information developed by those experts in the field is not Republican or Democratic; Catholic, Protestant, or Jewish; conservative or liberal. Whether or not the results of a given piece of scientific research support someone's preconceived notions, faiths, or beliefs, the results should be considered factual at least until subsequent studies prove otherwise. As someone who did scientific research for thirty-five years, I wouldn't be involved if it were otherwise. Everything I say in the rest of this book represents what I believe to be the consensus of opinions as discovered by the best researchers in the field and the teachings of the experts listed above.

3

In the Beginning

A N IMPORTANT aspect in assessing the current condition of the land is knowing what it looked like in the past. What is theoretically possible? What could it look like again? So what did the Hill Country look like in, say, 1836, when Texas won its independence from Mexico? Why choose that date? Because it is a time before there were large numbers of European settlers here, but still a time for which we have historical records of what the place looked like.

Before the nineteenth century, Texas was peopled primarily by Native Americans, with a smattering of Spanish missions and forts put up to convert the natives to Christianity and make them subjects of the king of Spain. Settlers of European origin were just beginning to trickle into East Texas and along the coast. Since the total population was quite low, and most of the Native Americans were largely nomadic hunter-gatherers, the influence of humans on the ecology of Texas, especially the Hill Country, was fairly low. The lifestyle of most Native Americans left what we would call today a small footprint on the land. Certainly, compared to everything that has happened since, that is true. So the first half of the nineteenth century is a good time to look back to and see what the land looked like before the heavy influences of humanity.

Fortunately, this same period saw the beginnings of visits by a series of educated individuals from the eastern United States and Europe. They were interested in the geology, flora, and fauna of Texas and collected literally tens of thousands of plant and animal specimens, which they shipped to research institutes and universities in the United States and Europe for study and identification. These early naturalists were also articulate note takers and letter writers. Thus we have what for the time were fairly accurate, scientific records of what these individuals observed.

This is not the place to discuss the various individuals who visited Texas, some of whom settled here, or their individual contributions to our knowledge of what nineteenth-century Texas looked like. But you will encounter many of their names in the common and scientific names of many plants and animals they discovered. Particularly noteworthy in this regard are Berlandier, Lindheimer, Roemer, Riddell, Wright, Bigelow, and Drummond. One interesting aspect of the collective reporting of these men (all of these early naturalists were men) is that reading each of their accounts is a little like the blind men describing an elephant: their description depended on which part of the elephant they felt—the "rope," the "snake," the "tree." For the most part the different naturalists saw different parts of Texas at different times. When they differ in their observations, it doesn't mean that one was right and the other wrong, but that they were all describing the incredible diversity of Texas.

Just to put things in historical perspective, Mexico declared its independence from Spain in 1821; the Texas Declaration of Independence was adopted on March 2, 1836. The siege of the Alamo ended on March 6, 1836; Santa Anna was defeated by Sam Houston at San Jacinto on April 12, 1836; and the Republic of Texas was recognized by the United States, France, and England, among others, in 1837. Texas was admitted to the Union on December 29, 1845. In 1842, German immigration into Texas was organized, leading to the establishment of New Braunfels, Fredericksburg, and other Hill Country towns. The population of Texas in 1850 was recorded as 212,000 (not including Native Americans).

I should explain what I mean by the "Hill Country." There is no hard boundary that one can draw and say to one side of that line is the Hill Country and to the other is some other ecological/political section of the state. Most people would define the south and east boundaries of the Hill Country to be more or less along the Balcones Escarpment, which runs roughly along I-35 from Austin to San Antonio and westward from San Antonio. The northern and western boundaries are less distinct, but they might be drawn roughly along a line from Lampasas to Mason to Junction to Rock Springs. The farther one gets outside that line, the more the ecosystem differs, especially as the soil and amount of rainfall change.

Geologically, the Hill Country could more accurately be called the "Canyon Country," because it started out as a flat seabed that was inundated by sea water repeatedly for millions of years, laying down multiple layers of limestone and limestone/clay deposits. Then, beginning sometime around the end of the Tertiary period (around 10 million years ago), this area was uplifted to form a flat plain that then began eroding as various creeks, streams, and rivers began making their way to the gulf. The depths of these resulting canyons or valleys vary, but most of the upland areas are relatively flat and for large sections of the Hill Country are at an elevation of around 2,000 feet. So early visitors coming from the coastal plain to the south and east would have observed what appeared to be hills rising before them as they approached the Balcones Escarpment and entered the Hill Country. On the other hand, if visitors had approached from the flat, higher areas to the north and west, they would have observed large expanses of flat land broken by increasingly numerous valleys and canyons carved by the many creeks and streams and their tributaries.

Most of the Hill Country is underlain by limestone. There are different formations of these limestones with different characteristics. One of the most interesting is the Edwards limestone, which is highly fractured and contains numerous small holes and larger voids (the term *karst* is used for these limestones), and it is this formation that makes up the famous Edwards Aquifer which is so vital to San Antonio. The part of the Hill Country that lies roughly north of the northern half of Gillespie County, called the Llano Uplift, is granitic in nature instead of the limestone found elsewhere.

We have all heard about the "tall grass prairie" that extended from Texas to Canada, where grass grew as high as a man's saddle horn in deep fertile soil. This was the goal of much of the westward migration as the United States expanded westward to occupy the area gained with the Louisiana Purchase in the nineteenth century. One of the things that made settlement of these prairies possible was the iron plow, because earlier wooden plows were no match for the tough roots of the tall grasses, making the breaking of ground for crops difficult. The Hill Country sits at the southern end of what was the tall grass prairie, and in fact much of the area may have been a mixture of tall grasses and mid-grasses, because the soil is thinner and the rainfall generally less than in areas to the north.

Putting together the accounts of all the early naturalists, one might conclude that the Hill Country contained a variety of vegetation types: dense woodlands, savannas, shrublands, and grasslands. In the canyons and stream and river valleys, woodlands of hardwoods and cedar probably predominated, possibly shading out the grasses in many areas, much as we see today. Cypress was seen along most rivers. On the flatter uplands there were areas of dense cedar stands as well as other areas with many hardwoods. But there were also many areas best described as savannas with small mottes of hardwoods or cedar and scattered isolated trees. And there were some areas of relatively treeless prairie or grassland. Many of the grasslands had significant stands of tall grasses.

An 1846 quote from Ferdinand Roemer illustrates the point:

On another occasion I made an excursion with Lindheimer to Mission Hill, which rises on the plain of the plateau lying north of the city [New Braunfels]. Our path led us again past the springs of Comal, but suddenly ascended the steep, wooded slope of the hill. . . . The cedar trees which covered

the slopes exclusively, formed an impenetrable thicket
through which a path had to be cut. . . . This cedar forest
was a treasure to the colonists of New Braunfels, since the
wood is preferred above all others on account of its durabil-
ity when used in building houses and fences. A section of
this cedar forest was destroyed by a forest fire during my stay
in New Braunfels. . . . As soon as we reached the summit of
the hill, the cedar forest ended. An open grassy plain, only
broken here and there by brushwood and scattered live oaks
trees, spread out before us. It extended to Mission Hill about
two miles distant and we had to follow a narrow Indian trail
to reach it.

We are all familiar with the fact that there used to be huge herds
of bison roaming the prairies from southern Texas to Canada. They
moved in very large herds that left little in the way of edible forage
behind. But their movements were somewhat random, probably
dictated by combinations of factors such as wind direction, quality
of the grass, prior memory of certain areas, and in some areas geological features such as river crossings and slopes. The bottom line,
however, was that the bison did not normally visit the same area
more than once every several months, and more often than not
they might not come back to the same place for a year or longer.
This behavior allowed the grass to recover from being grazed and to
grow back and set seed. So, although the immediate impact of the
bison was to graze down the grasses, their presence did not significantly reduce the productivity of the prairie, and in some cases the
bison may have contributed to the general health of the savannas
and prairies. The bison were essentially eliminated from the Hill
Country by about 1850 and from all of Texas by about 1880.

Another feature of the area in the early 1800s that might be less
familiar is the frequency of fire. Imagine if you will large expanses
of prairie composed mainly of dense stands of grasses two, three,
four feet and higher. Now imagine a dry summer where the grasses
go dormant and dry out. Then a lightning storm ignites the dry

The bison is but a symbol of what this land used to be like, along with all the other animals, grasses, forbs, and trees that made up the Hill Country in the early nineteenth century.

grass and the wind carries the fire across the savanna. What will cause it to stop? Maybe a rainstorm, but the probability of that is low. Maybe when it runs out of fuel, perhaps by burning up to a stream or river or maybe to a deep valley where the trees have shaded out the grass. The point is that in the hot, dry, windy summer, there is not much to stop a wildfire, and large areas can be burned from a single lightning strike, or a Native American campfire spark, or a fire intentionally set by Native Americans.

It turns out that after a fire, when the grass sends up fresh, tender shoots, grazing animals love it. The natives knew this and sometimes set fires to attract game. Also, if the Comanches were attacking the Apaches, the latter might set a fire to slow down their attackers. The point is that fires happened with some frequency. Most biologists and range scientists today believe fire occurred in most places every five to seven years on average. Ferdinand Roemer mentions fire five or six times in his book chronicling his eighteen months in Texas. Here's one interesting example, on a trip from New Braunfels to Fredericksburg:

> During the night a prairie fire caused us considerable worry since it approached us from several directions. We resorted to the usual method of protection by burning the grass for a certain distance around our camp. Not until we had thus assured our safety, could we enjoy the beautiful spectacle. In the darkness of the night the strips of fire, several miles in extent, appeared as fiery brands, which, governed by the strength of the wind, moved forward now quickly, now slowly; the flames shooting up high or just glimmering, according to the length of the grass. An especially beautiful view was afforded by a group of live oaks, on which magical illuminations were cast by the burning grass. . . . These fires often cause severe losses near the settlements, since they destroy the fences enclosing the fields. When settlers sight a prairie fire nearing their farms, especially during a strong wind, all the inhabitants hurry to the threatened place, and an effort is made to extinguish the flames by beating them with wet sacks, or a ditch is drawn to halt their advance.

To summarize, in the early 1800s the Hill Country would have looked much like it does today. The flatter uplands would have probably had more areas of grassland with fewer trees than we have today. These grasslands were grazed heavily by migrating bison, but they had long rest periods with little or no grazing. Grassland fires burned the prairies with some regularity. Overall, the percentage of land covered by cedar and live oaks was less than we see today, and the grasslands contained more grasses and forbs and a greater diversity than is typical today. We are, in fact, lucky to live in an area as little changed as the Hill Country; many areas of this country are vastly changed from their earlier condition.

The consequences of the grazing bison and the occasional fire actually helped make the prairie extremely productive in terms of both the total biomass produced per year and the biodiversity of the grassland ecosystem. It has taken most of us more than a century to appreciate these facts, and to appreciate how good a model the nineteenth-century prairies and savannas are for twenty-first-century ranches.

Although the most obvious changes in appearance of the Hill Country from the early nineteenth century to today have to do with the explosion of the human population and the changing plant communities, animal makeup of the area is also different. Early nineteenth-century Hill Country saw black bear, wolf, mountain lion, bobcat, ocelot, coyote, alligator, bison, pronghorn antelope, deer, and javelina. Of this list, the black bear, wolf, ocelot, alligator, bison, and pronghorn have been essentially eliminated from the Hill Country, mountain lions and bobcats are uncommon, as are javelina in most of the area. Only deer and coyotes from this list are in any way common in today's Hill Country, although most of the others are common in other parts of the state. Black bears exist in isolated numbers in East Texas along the Louisiana border and are coming back in the Big Bend area, and ocelots are limited to a few counties in the Lower Rio Grande Valley.

Bison were common even in south Texas in the early nineteenth century, but they were mostly gone from central Texas by 1850 and from all of Texas by the 1880s. The last recorded black bear was shot in Kerr County in 1909, and the last wolf was taken in 1913. Because prairies are not particularly good deer habitats, these browsers were more concentrated in the woodlands of the canyons and valleys, leaving the prairies to the bison and pronghorn. Interestingly, hunting pressure from settlers as well as Native Americans reduced deer populations in areas around settlements significantly by the late 1840s.

One final point about what the land looked like in the nineteenth century: In areas where management practices have best simulated short-duration grazing followed by long rest periods with no grazing, plus occasional prescribed fires, the land begins to resemble nineteenth-century Hill Country more closely. In those areas, productivity of the land has increased considerably, coming closer to our nineteenth-century model. So when we talk about what this land should naturally look like, we don't have to rely on just historical accounts. We can look at the character of the better-managed ranches in the Hill Country and see what the land does look like under conditions of lower stocking rates, fewer deer, and occasional fire.

4

What Have We Done?

Now that we have some idea of what the Hill Country was like before significant numbers of European settlers arrived, we can begin to understand the effects these people, our ancestors or predecessors, had on the land. Some of the changes caused by the settlers will not be viewed by our twenty-first-century eyes favorably, but it is not my aim to criticize what these people did or to blame anyone who came before us for the conditions we find today. In fact, the more I learn about what life was like in the latter half of the nineteenth century or even the early part of the twentieth, the more respect I have for those folks; they endured some severe hardships just trying to survive and make a living. In addition, our general knowledge of the Hill Country ecology and of the complex ecosystems therein has changed significantly in the past forty or fifty years, so what was considered a good thing to do in the past may now be viewed differently.

The settlement of Texas began slowly in the 1820s, with the first Anglo birth in Texas believed to be in 1821. Settlement increased in the 1830s and became really significant in the 1840s. After statehood, the rate increased even more. The first census, in 1850, found 212,000 people in the state: 154,000 whites and 58,000 blacks, many of whom were slaves; Native Americans were not counted at this time. Even with this increased immigration, 212,000 people in 170 million acres would have meant only one person for every 800 acres overall. But since many folks lived in Galveston, Houston, San Antonio, and other villages, the actual population density in the rural areas was even less. So even in 1850, the number of settlers of European ancestry hardly seems to have been numerous enough to change the landscape radically, but the beginnings of this change were certainly occurring.

As the settlers moved into Texas and spread out across the land, they brought with them not only their European-style society and agriculture but also their exotic animals—what we now call cattle, sheep, goats, pigs, and chickens. Horses had been introduced earlier by the Spanish as they traveled into Texas from Mexico. (Note that I am not differentiating here between settlers coming directly from Europe and those who were first-, second- or third-generation Europeans migrating from various parts of the United States, but mostly from the southern states.)

As an aside, it is interesting to note that the vast majority of the world's domesticated livestock are of European, Asian, or African origin. The Western Hemisphere simply did not contain many useful, domesticatable animals. This fact has been used to explain why Old World societies were more advanced than those in the New World: because the former had domesticated livestock, they could spend less time in the pursuit of food and thus stay in one place rather than constantly roam in search of food. This allowed them the luxury of living in cities surrounded by many more people from whom they could learn, and also to have the time to spend in other pursuits such as building, learning, thinking, and traveling.

Whether this is true is debatable, but what is not in dispute is that Europeans had a much bigger effect on the Hill Country ecology than did Native Americans. This is explained partly by their numbers and technology (horses, guns, etc.) and partly by their lifestyles and animals (permanent settlements, grazing the same area constantly, and clearing and farming some areas).

This circa 1869 farmhouse and barn at the Lyndon B. Johnson State and National Park give us a glimpse of what life was like for Hill Country residents in the latter half of the nineteenth and early twentieth centuries.

The Native Americans essentially had a subsistence lifestyle; they took only what they needed from the land for their own use, and for the most part they traveled from one place to another frequently, so the effects of their activities on any one area were temporary. Some tribes did indeed make semipermanent settlements and engage in some farming, but the vast majority were hunter-gatherers. There are some reports of natives deliberately setting fires, and they surely caused some accidently, all of which contributed to the lack of trees on the plains.

By the time most settlers arrived in the Hill Country, the bison population had been greatly reduced. Bands of hunters had earlier killed most of the bison and many deer and antelope for their hides, which were shipped back to the East Coast.

As settlers moved in, they built houses, sheds, and various other structures; fenced off gardens and farm fields; and allowed their animals to roam freely nearby. Eventually, as the density of settlers increased and especially as barbed wire became available, they also fenced in their livestock. The most important change this brought was the continuous grazing of grasslands, which altered both the quantity and quality of the forage; the better grasses began to disappear, and smaller, less-palatable grasses began to increase. The rate of these changes depended largely on the number and type of animals.

The settlers depended heavily on their livestock, and their standard of living depended largely on how many animals they could raise, both for their own consumption and to exchange for essential goods. The loss of a lamb or a calf or even a chicken was a major setback for the whole family, so the protection of these animals was important. Predators were routinely shot at every opportunity, as were the game animals that made up a significant part of early settlers' food supply.

Similarly, fire was a great threat that had to be fought vigorously. Not only could fire destroy homes, sheds, storage buildings, and fences, it also destroyed animal feed. Settlers depended on the grass year round to feed most of their animals, and a grass fire could leave them with little or no forage for several months. Fires probably became somewhat less extensive as settlements increased, because the smaller fuel load of shorter, more heavily grazed pastures would not have carried a fire as quickly or as far as a fire in a tall grass prairie.

As a consequence of the practices of these new human inhabitants, the grasses became shorter and fires became less common over any given area. This allowed woody plants and forbs of all kinds to gain a foothold in areas that previously had many fewer trees. These changes in vegetative composition, in addition to the reduced numbers of predators, meant a much improved deer habitat.

The settlers undoubtedly noticed the reduction in the amount of grass in areas where their animals were grazing, as compared to areas that had not yet been settled or grazed. But they had no way of knowing the long-term consequences, and in any event they had to worry about surviving the current year. Their knowledge of raising farm animals came from their experiences in Northern Europe or the East Coast—all areas of higher rainfall and deeper soil. What they were not prepared for was the high frequency of significant drought in this area. The Hill Country lies between areas of significantly higher rainfall to the east and significantly lower rainfall to the west and experiences periods of higher- and lower-than-normal rainfall with equal probability.

In fact, many settlers didn't recognize the latter point and assumed that the higher-than-normal rain years were normal and that drought was unusual, abnormal. This distorted view of the Hill Country climate persisted with some ranchers well into the middle of the twentieth century and contributed to the management style of always stocking pastures on the assumption of a good rainfall year, inevitability leading to overgrazing.

Throughout the second half of the nineteenth century, the population of both people and livestock increased considerably as settlements grew into villages and villages became cities and the "frontier" was moved farther and farther west. As these changes were taking place, the effect of more people and livestock on the Hill Country ecology became even more pronounced.

Ranchers in much of the early twentieth century, before World War II, had to contend with drought, depression, and other economic hardships. These conditions led most ranchers to concentrate on surviving another year rather than on the long-term productivity of their ranches. This mindset in turn led to general overgrazing of the land and a gradual reduction in the productivity of Texas ranchland. Although most of the large predators had been eliminated or severely reduced in numbers by the early twentieth century, among the other problems ranchers faced was the screw-worm fly. This fly laid its eggs in small cuts or scrapes in an animal's skin, and the larva ate at the flesh, causing large wounds, infections, and death. In 1935, Texas ranchers lost an estimated 180,000 head of livestock to the screw-worm fly. As a boy in the '50s raising 4-H animals, I was told to be very careful to watch for injuries and doctor every cut.

Research conducted at the USDA insect research laboratory near Kerrville eventually led to a method of controlling the screw-worm fly population, and by the early to mid-1960s the pest was essentially eliminated from Texas. (It has still not been eliminated from Mexico.) We don't know the numbers, but one can imagine that if livestock, which were being looked after and cared for, had such large losses, then the white-tailed deer population had even greater losses. With the elimination of the screw-worm fly, the last real predator of white-tailed deer, other than humans, was gone, leading to a further increase in the deer population from the 1960s into the 1990s. The first statewide deer census, in 1935, estimated a population of about a half million. In the 1990s, the population was estimated to be around 3.5 million. It was in this period that the deer population began to reach levels capable of severely impacting the rate of replacement hardwoods in the Hill Country, which is why we see many hardwood trees that are thirty to forty years old or older but many fewer that are two, ten, or twenty years old. On many properties in the Hill Country, there are no small saplings (trunk diameter 1–3 inches) of any of the hardwoods unless they are growing along a creek bank out of the reach of deer, or maybe inside a large prickly pear or agarita bush.

The most notable exception to the decrease in trees or shrubs because of the lack of replacement is cedar (*Ashe juniperus*). Deer have favorite foods, less favorite foods, unfavored foods, and things they simply will not eat. Cedar falls farther down the deer preference list than most of the hardwoods. So, though deer repeatedly eat the leaves off most hardwood plants they can reach, they usually leave a cedar alone. Some of the other plant species that have increased are Texas persimmon, Texas mountain laurel, whitebrush, and Mexican buckeye—all less favorite deer foods. Almost all livestock and exotic ungulates have similar preferences. Goats are known to eat small amounts of young cedar, but their digestive system can tolerate only so much, and even then they prefer many other foods over cedar if they can find them.

What kept cedar from being far more pervasive in the 1800s were the fires that raged across the prairie periodically. But now that fires are rare, and nothing eats the young cedar bushes, there is little if anything to keep the numbers in check, other than landowners.

So what is wrong with cedar? I discuss this in detail in chapter 9, but the simple answer is that cedar tends to crowd out other plants including grasses, thus reducing biodiversity and the ability of the land to sustain many different species of plants and animals. Additionally, the role of cedar in intercepting rainfall and affecting stream flow is an ongoing subject of study and one that engenders different points of view. In some instances, removal of cedar seems to have resulted in measurable increases in spring and seep flows. As the population of the Hill Country increases, a full understanding of the role of cedar in rainwater capture will be even more important.

5

What's the Problem?

ERE we are in the beginning of the twenty-first century, looking back at what has happened to the Hill Country since European settlers began arriving six, seven, eight generations ago. People still think of the Hill Country as one of the most beautiful parts of the state; they still flock to the area to buy a little piece of "paradise" and retire here. In fact, it seems to me that people are by and large very happy to be here and consider the Hill Country the best possible place to live, something one can't say about many places anymore.

So, from the standpoint of the environment, the ecology, and the landscape, what is wrong with the Hill Country? As a naturalist, I see eight main ecological problems: overgrazing, overbrowsing, cedar encroachment, erosion, reduced water availability, oak wilt, exotics of all kinds, and the increasing human population. Some of these problems may be more severe in one place than another, but collectively the Hill Country suffers from all of the above.

It is important to note that these are not isolated problems; they are related. Most are exacerbated by increasing human populations and the associated land fragmentation, and most (except for the population increase) can be at least somewhat mitigated by good land management practices. Each of these problems is discussed in more detail in later chapters.

Overgrazing. Overgrazing can be defined as having more grazing animals on land than the carrying capacity of that land. Carrying capacity is the density of animals the land can support *without degrading the habitat.* Put another way, a given piece of land can produce only so many pounds of edible forage in a year. If the number of animals on the land is such that they don't eat all the forage produced in a growing season, then the area is not being overgrazed. But if the number of animals is such that they eat all the land produces and require more, then they are degrading the habitat and the land is overgrazed.

Overgrazing does several things to the land. It reduces the size of the grass plants, and smaller grass plants produce less forage, thereby leading to a reduction in the amount of grass the land can produce. It leads to an increase in the amount of bare ground, resulting in an increase in soil temperature and a decrease in rainwater infiltration rate, an increase in runoff, and an increase in erosion. And it changes the distribution of grass species from one containing more mid- and tall grasses to one of more short grasses, from one of many palatable grasses to one of many unpalatable grasses. The overall biodiversity of plants and animals is less in an overgrazed landscape, where there is a reduction in both foodstuffs and habitat for many small animals.

Overbrowsing. Overbrowsing can be defined as having more browsing animals on the land than the carrying capacity. A cow normally consumes about 85–90 percent grass and 10–15 percent or less of forbs and woody plants. A white-tailed deer normally consumes about 85 percent woody plants and forbs and 15 percent or less of grass. Thus, cattle can be considered complete grazers and white-tailed deer complete browsers according to the above definitions. Sheep, goats, and most exotics have diets that put them between white-tailed deer and cattle; that is, they all eat significant amounts of grass, forbs, and woodies, at least over the course of a year. The carrying capacity for browsers is the density of browsers the land can support without degrading the habitat.

The lack of any edible vegetation below the browse line, typical of much of the Hill Country, is the result of the overpopulation of deer (see photo on facing page).

Overbrowsing has caused a general decline in the number and variety of most woody species in the Hill Country. It is preventing the growth of most small trees and shrubs, thus eliminating the reproduction of these species and causing the decline in woody plants in the Hill Country. Overbrowsing removes food, cover, and general wildlife habitat from the land. We have probably seen the highest populations of trees and shrubs in the Hill Country and are now seeing a decline.

Cedar encroachment. We have few fires to keep new growth in check, and, in contrast to the hardwoods, nothing really likes to eat

In areas of low deer population, or areas fenced off from the deer, as in this photo, tree limbs grow down to the ground and low shrubs give rise to a substantially improved habitat. The trees in this and the preceding photo are blackjack oaks, growing only fifty feet apart. Those in the previous photo are accessible to deer, while those in this photo are not.

cedar. Therefore, nothing controls the growth of cedar other than humans; there is no effective biological control. The consequences are that cedar trees will continue to grow larger and more numerous just about anywhere in the Hill Country unless we intervene. Because the unchecked growth of cedar leads to the crowding out of other plants, biologically diverse areas containing numerous forbs, grasses, and woody plants become converted to a cedar monoculture (a solid stand of only one species) that supports very few animals. Furthermore, although there is disagreement among experts and some evidence to the contrary, cedars in certain situations may intercept more water than hardwoods or grasses, decreasing the amount of rainwater reaching the ground to nourish plants and recharge springs and seep aquifers.

Erosion. The most important thing we have on the land is the soil. Without soil, we can't grow plants, and without plants we won't have any animals. The regeneration of soil by the natural weathering of rocks takes hundreds or thousands of years, so whatever soil you have on your land now is about all you will ever have. The problem is that, if erosion is allowed to occur, you may lose the little soil you have naturally. Erosion can occur because people have disrupted the landscape, creating new drainage channels and exposing new bare dirt, by our road- and house-building activities, by mechanical brush-clearing activities, or by various other farming activities. Erosion can also occur because of livestock activities: overgrazing that reduces the amount of grass cover and increases the amount of bare ground; trampling where animals spend a lot of time, such as in areas where supplemental feed is provided and especially around water sources. Erosion of riparian areas (land near lakes and streams) is especially detrimental because it results in lost soil at the point of origin and pollutes the water.

Water availability. It is not uncommon for municipalities to have to institute water restrictions during dry summer months. It is also not uncommon for individuals living in the country to have their water wells fail because of drops in the water table. Part of this problem obviously is caused by the increase in population, which increases the demand for water. And there is no doubt that much of that demand is for water-wasting or unnecessary activities. We are not likely to be able to do anything about the population increase, but education may well help us all use water more efficiently and less wastefully.

Another reason our water supply does not meet the demand may be that we are not capturing our rainwater effectively. And here I am not talking about rainwater harvesting from roofs of buildings, as admirable and practical as that is. I am talking about land management practices that fail to capture rainwater efficiently.

Developers sometimes do destructive things to the landscape in the name of progress.

Overgrazed pastures can have excessive runoff, and a landscape covered with cedar may intercept or use water in a different way than land covered in native grasses. Land management affects both the quantity and the quality of our water.

Oak wilt. This disease has become a serious problem in the eastern two-thirds of the Hill Country. It is caused by a fungus that attacks oaks, primarily red oaks and live oaks, and plugs up their vascular systems. The trees literally wilt from lack of water and nutrients being transported to the leaves. This is similar to the Dutch elm disease that decimated elm trees in much of the country many years ago. It is especially troubling in the Hill Country, not because the disease is more virulent but because of the types and distribution of trees here. For much of the Hill Country, live oaks are by far the most numerous hardwoods and almost the only hardwood growing on some properties. Live oaks tend to be connected to their live oak neighbors by their roots, either because they are root sprouts of the same tree or because their roots have grafted together underground. Whenever a live oak contracts the disease, the fungus can travel throughout the root system and infect neighboring live oaks, thus killing all or most of the trees in a given area. Blackjack and Spanish oaks (Texas red oak) are also affected by oak wilt, but this usually only involves isolated trees, leaving neighboring trees unaffected. Oak wilt is obviously accelerating the general decline in the number of hardwoods in the Hill Country, and it remains to be seen how significant this problem will become. For those who lose prized trees around their homes, the disease is already devastating.

Exotics. The early European settlers were the first to bring exotic animals into Texas, the ones we now call cattle, horses, sheep, goats, pigs, and chickens. Beginning in the mid-twentieth century, Texans began to import all manner of exotic animals from Africa, Asia, and Europe, mainly ungulates. Of course, these are not the only exotic species we have to contend with now. We have fire ants, Formosan termites, Africanized bees, zebra mussels, nutria, feral hogs, and feral cats. Plants too: Chinese tallow, chinaberry, Chinese pistache, bamboo, salt cedar, KR bluestem, johnsongrass, ligustrum, vitex, kudzu, nandina, and water hyacinth.

Some of these and many, many more were brought here intentionally, some accidentally, and some we don't know how they got here. The problem posed by many of these exotics is that their growth or reproduction is uncontrolled to the extent they begin to crowd out native species and alter the native habitat. Many, even most of the plants and animals brought into this country cause no serious problems, mainly because they do not or cannot escape cultivation or captivity to propagate on their own. But a small percentage do just that. For the most part, the exotic ungulates that were intentionally brought here are generally confined to the owner's property. An exception is the feral hog, which has been free roaming for many years and is very destructive. More recently, axis deer have become free ranging in some areas. The main problem with these animals is that they eat much like goats do, which is to say they eat browse (woody plant leaves), forbs, and grass and can survive on any one of them, putting them in competition with white-tailed deer as well as livestock.

Land fragmentation. Like most other problems we face in twenty-first-century America, the ecological problems of the Hill Country are exacerbated by increasing populations of humans. The more people we have, the more electricity and water they demand, the more sewage they generate, and the more roads, buildings, parking lots, and other impermeable surfaces they construct, thus causing more rainwater runoff. More people means greater use of chemicals, the introduction of more exotic plants and animals, more parcels of land poorly managed, as well as more free-ranging pets and feral animals. One of the worst consequences of increased populations is a greater demand for small parcels of land, which puts pressure on larger landowners to subdivide their land. This fragmentation results in more fences and a patchwork of habitats that in general is much less beneficial to wildlife and livestock than the original larger acreages.

Although good land management practices can mitigate almost all of the problems discussed so far in this book, what they can't do is limit population growth. This fact makes it even more important that everyone adopt good land management practices. I have enumerated a long list of problems here, but, lest the reader become discouraged, there is also a growing awareness of these problems and a greater willingness to fix them now than there has ever been. The financial ability of new landowners to implement solutions to the problems is also greater than in the past.

In the next few chapters we explore each of these problems, as well as several other issues, in more detail, along with the solutions or preventions that can be applied to improve and protect our Hill Country.

6

Every Case Is Different

B EFORE we start discussing how to manage your property
to best mitigate some of the potential ecological problems,
it is important to make one point about any advice you
receive on managing your place. There is little about land management or biology itself that is absolute, that applies to every situation under every circumstance at all times.

When asked a question about any of the topics they are most
expert in, virtually all of the experts listed in chapter 2 will begin the answer with "It depends. . . ." Two that have been most
insistent that answers to ecological, land management questions
begin with that phrase are Bill Armstrong, retired wildlife biologist
with TPWD at Kerr Wildlife Management Area, and Lew Hunnicutt, former faculty member at Texas State University and former
education director at the Bamberger Ranch Preserve.

When discussing a certain condition found in the Hill Country,
or especially a suggested practice to mitigate a land management
problem, it is impossible to describe a practice that applies equally
well to all situations on all properties. Almost nothing in this area
is all black or white, almost nothing works all the time, and almost
no condition occurs on all properties in the same way. Responsible experts almost always make that point. If someone expresses

Cartoon of Lew Hunnicutt, then education director at the Bamberger Ranch Preserve, emphasizing overstocking as one of the main problems in the Hill Country landscape, along with one of his favorite phrases: "It depends."

a high degree of certainty about what you should do on your property, especially without seeing it, take that advice with a grain of salt—and that goes for what you read in this book too. What works in most places most of the time may not work for you. You should listen to the discussion and the explanations, but don't blindly follow any advice unless it makes sense to you.

You have to make a judgment about whether a particular problem exists on your land and whether a suggested remedy will work in your particular situation. To do that, it is important that you understand the basic causes of the problem and how the suggested solution works to address those causes. The discussions in the following chapters are intended to help you do that.

7

Overgrazing

BEFORE we can discuss overgrazing in detail, we should review some of the basics of range science, animal behavior, and plant growth. I call this Range Science 101.

Food Selectivity

Just as we are selective in the things we like to eat and the things we can't or won't eat, so are livestock. Different species have different preferences, and there are even some differences among individual animals. These different preferences are largely based on the biology of the animal, which determines which foods it can digest and which have the proper nutrition for it to grow and live. But what an animal eats is also determined by what is available, just as we might prefer sirloin steak and chocolate cake, but if only hamburger and Oreo cookies are available we won't turn them down.

Research done at the Kerr Wildlife Management Area and elsewhere has determined the food preferences of most livestock species as well as many exotic animals and white-tailed deer. It probably won't come as a surprise that cattle eat grass, but you may not be aware that they also eat up to 15 percent browse and forbs. Horses are stricter grass eaters; sheep like mostly grass but eat more browse and forbs than cattle; goats eat roughly equal amounts of browse, forbs, and grass, if available; and most exotic ungulates (hoofed herbivore) eat like goats. White-tailed deer eat almost exclusively browse and forbs, with at most only about 10–15 percent grass.

Most of these animals are ruminants, meaning that they have a four-compartment stomach where the food they eat is first broken down by bacteria in the front part of the digestive system, the rumen, before being absorbed into the bloodstream to nourish the animals. Horses are not ruminants (they don't chew cud), and their food is broken down by bacteria in the hind part of their digestive system, called the cecum.

Grass, tree leaves, and forbs are all primarily cellulose. Humans couldn't digest any of these; if we were to eat them, it would be indigestible "fiber," because we don't have the bacteria capable of breaking down cellulose. We can digest only the parts of the plant that are starches, sugars, proteins, or oils. Starches and sugars are related to cellulose (they are all carbohydrates), but the former are much easier to break down into small molecules that can be absorbed into our bloodstream, so we don't need the same bacteria in our system that ruminants do.

It turns out that not all cellulose is the same either. Some cellulose sources (grass) contain a lot of lignin, which makes their digestion even more difficult; other sources (tree leaves and forbs) contain less lignin and are easier to digest. Wood (tree stems) contains even more lignin and not even cows can eat wood (although termites can).

Some of what animals choose to eat is determined by their biology. Cattle have relatively large rumens which are highly compartmentalized, giving the food a long residence time for the bacteria to break it down. Cattle have large mouths and long tongues capable of wrapping around a bunch of grass and pulling it off. They are not well suited for picking small individual leaves off tree branches or picking up small forbs growing amongst the grass. Sheep, goats, most exotics, and deer have small mouths with flexible lips capable of very selectively eating just the parts of the plants they want.

Animals have preferences even for some species of grass over others; some browse plants are much more highly favored than others. One might suppose that animals would prefer tender, soft plants over those that are coarse, rough, or have thorns or stickers. Although this may be a minor factor in the animal's choice, it is clearly not the main factor; not all soft tender plants are eaten, and certainly many plants with prickly, sharp, or pointy parts are readily eaten, even favored. Clearly, taste and smell must play a part in determining if a given plant is highly favored or not.

Grazable Acres

Cattle almost never utilize all of a given pasture equally. (Note: I use the term pasture here to mean a given fenced section of a ranch of native, unimproved grassland. I am not referring to irrigated/cultivated flat pastures of coastal bermudagrass, kleingrass, or sudan of a few hundred acres; cattle may well graze those pastures quite uniformly.) In the Hill Country, cattle spend little time more than a mile from water, they spend almost no time in dense cedar thickets (there is usually no grass there, anyway), and they tend to avoid very rocky areas or very steep slopes. And, of course, areas within the pasture that the animals are fenced out of, such as around homes, gardens, roads, and driveways, don't provide any forage for them either. The point is that not every acre inside your fence is necessarily grazable. It is quite possible, especially where there is a substantial amount of cedar, for the number of grazable acres to be less than half the total acreage within a fence. When determining how many animals you can raise on a given property, it is essential that you calculate using the actual grazable acreage.

It is not at all uncommon for someone, having spent most of their life in the suburbs, to move to the Hill Country on 100 acres and think they can raise large numbers of animals on this huge

A severely overgrazed and overbrowsed pasture. This scene is all too familiar in the Hill Country, but is slowly becoming less common.

"spread." They are surprised to learn that—because the area was recently overgrazed, 30 percent is under cedar, and the hill up to their house is steep and rocky—even two cows may be too many for that land to support.

Carrying Capacity and Animal Units

The **carrying capacity** of a given piece of property is the density of animals the land can provide food, water, and shelter for *without degrading the habitat.* In the Hill Country, carrying capacity is often described in terms of acres per animal unit. Conceptually, it is a simple idea. With a given amount of rainfall, a certain piece of land is capable of producing only so many pounds of forage a year. In good rainfall years that amount is higher than in poor rainfall years.

An **animal unit** is defined as a 1,000-lb cow (plus a calf with her for six months of the year), and it is expected that this animal unit will consume 26 lbs of dry forage per day. The number of animals of other species that it takes to eat 26 lbs of dry forage a day also equals one animal unit (see table). It is useful to think in terms of animal units because it allows you to estimate how much different numbers of different sizes of animals will eat in a given day. It turns out that 26 lbs of dry forage a day works out to just a little less than 10,000 lbs a year.

Typical carrying capacities for native pastures in the Hill Country range from 20 to 30 acres per animal unit. These numbers need some explanation because they are sometimes a source of confusion. A carrying capacity of 30 acres per animal unit is a *lower* carrying capacity than one of 20 acres per animal unit. If one has 300 acres with a carrying capacity of 30 acres per animal unit, then the land can support ten mother cows, but if the carrying capacity is 20 acres per animal unit, the land can support fifteen mother cows.

Kind of Animal	Animal Unit Equivalent
Cow with calf	1.00
Bull	1.35
Horse	1.25
Sheep	0.20
Goat	0.15
White-tailed deer	0.15

Stocking Rate

The **stocking rate** is usually described as the number of acres per animal unit that you have on your land. (Note, that as discussed above, 30 acres per animal unit is a *lower* stocking rate than 20 acres per animal unit.) If the stocking rate is less than the carrying capacity, then the animals have more than they need to eat and the habitat will not be degraded; in theory, you can continue to have that many animals on the land well into the future without any problems. If the stocking rate is greater than the carrying capacity, then the animals are eating more than the land can produce on a continuing basis, and therefore the habitat will be degraded, become progressively less productive, and the animals may be malnourished as well.

How Grass Grows

If you could reach down and get a hold of a large perennial native bunchgrass and pull it totally out of the ground with all of the roots still intact, you would see that the part of the plant below the ground looks much like the part above the ground, and that in fact the total mass of the root system is at least as great as the green part above the ground. The long fibrous root system of a native perennial grass is not branched like the branches and roots of a tree or forb; instead, all the roots originate near the surface of the ground (just as most grass blades do). Associated with the root system is a whole community of organisms, from beetles and earthworms down to bacteria and fungi. The roots exude carbohydrates on which these organisms feed, and in turn the organisms fix nitrogen into a form absorbable by the roots and make other nutrients soluble. About a third of these roots die every year and are replaced in the normal course of the life of a grass. As the dying roots decay and feed the microorganisms, they also leave behind the "tunnels" that the roots once occupied. These, plus the tunnels formed by some of the larger animals living among the roots, make the soil under the grass plant very porous, much like a sponge, and this is the reason water soaks into the ground under a grass plant much faster than into bare soil.

In the normal life of a grass plant, the roots absorb water, minerals, and nitrogen (in the form of nitrogen compounds, not elemental nitrogen), and these substances are transported to the leaves through the vascular system of the plant (similar to our blood vessels). In the leaves, CO_2 from the air is absorbed and reacts with water to form carbohydrates by the process called photosynthesis, which captures energy from sunlight to carry out that reaction. Some of those carbohydrates are transported back down to the roots for root growth and replacement, and the rest become more leaves. Thus, it takes roots to make leaves and leaves to make roots.

What happens when a cow comes along and pulls off some of the leaves of this grass plant? The remaining leaves continue to photosynthesize, creating more carbohydrates that ultimately become new leaves and replacement roots (to replace some of those that died naturally). At least, that is true up to a point. As long as the amount of leaves taken by the cow is half or less of the total leaves, the plant has plenty of leaf surface to carry out enough photosynthesis to restore the plant to its original state. But if the cow takes more than about half of the original leaf area, or if another cow comes along before the grass plant can recover from the first bite and takes even more of the leaves, leaving only a small amount remaining, then the plant does not have enough surface area to carry out enough photosynthesis to restore the plant to its original condition. Some of the roots will not be replaced, and the result is a smaller, less healthy, less productive grass plant.

A small grass plant cannot make as much new growth during the next growing season as a large grass plant. A grass plant with fewer roots cannot make as much new growth as one with many roots. It turns out that, on average, a grass plant that is only half or less grazed can regrow to its original size during the next growing season. But a plant that is grazed more than that, leaving less than half, will be weaker and less productive and will not be able to grow back quite as strong as it was originally. Less forage will be produced next year than would have been the case if grazing had taken only half or less. Thus one of the more important range science sayings: "Take half, leave half."

Now let's think about a pasture that has two different grass species, one the cows like and one they don't like. The rancher, not taking this fact into account, assumes there is enough grass to feed a certain number of animals and puts that many in the pasture. Cattle, being selective eaters, go after their favorite grass (call it "yum") and leave the other one alone (call it "yuck"). Pretty soon all the yum plants have been grazed once, but the cows are still hungry, so they graze on yum some more. The result is that the yum plants will be weaker and smaller than they were, but the yuck plants will be unchanged. And if that same practice occurs year after year, pretty soon there will be many fewer yum plants in the pasture, and the yuck plants will be even more plentiful because they can thrive without the competition from the yum plants.

Yum plants are called *decreasers* because they decrease with overgrazing, whereas yuck plants are called *increasers* because they increase with overgrazing.

Productivity

To put all of these factors together into a coherent picture, we have to begin with the most important component of your land—the soil. Your property, or maybe each ecological area of your property, has certain characteristics that are pretty much out of your control: you live at a certain fixed place on the planet (latitude/longitude), which means that you have certain climatic conditions—rainfall, temperature, length of daylight, frost-free days. In addition you have certain soil properties—soil depth, texture, composition, pH, cation exchange capacity—as well as geological features—slope, springs, seeps, orientation, rockiness. All of these things determine the potential productivity of your land in terms of the number of pounds of forage per acre per year you can grow under ideal conditions. In most of the Hill Country, this number is between 2,500 and 4,500 lbs/acre (the larger numbers are more likely in the east, where rainfall is greater).

To repeat, we are talking about the *potential* productivity under ideal conditions. Ideal means average rainfall total and distribution throughout the year, good soil condition, and good stands of the kind of grass that would normally be growing in your area if it were not grazed. Range scientists categorize rangelands in terms of how closely the condition of the land matches the historical climax condition (potential productivity) that would have been expected for this land. Land that is between 75 and 100 percent of that condition is considered "excellent," and very few Hill Country properties fit that category. Ranges that are in lesser condition are given lesser scores, and the potential productivity of each of these is correspondingly less.

Several things reduce the potential productivity of the land. Trees are one; they may be a positive contribution for browsers, but they are negative for grazers because they shade the pasture, reduce grass growth, and capture some moisture. Cedar brakes are the worst. Other things that contribute to lower potential productivity include overgrazed conditions with less than the expected amount of grass and more bare ground, less than optimal species composition, erosion reducing the amount of soil, and man-made structures such as buildings, roads, and ponds.

So, if your land were in excellent climax condition with the corresponding amounts and composition of grasses, it might be expected to produce 4,000 lbs/acre of grass per year. But if your grass is shorter because of past overgrazing, you have less of the expected good grazing grasses, some soil has been lost, and you have some thick oak groves and cedar brakes, then the condition of your land might be only fair; it might be expected to produce only 2,000 lbs/acre of grass in the open areas, and the areas under the oak groves might produce only 500 lbs and under cedar essentially zero. If the area under the oak trees amounts to 15 percent and the area under cedar 35 percent, then your range will be capable of producing only 1,075 lbs/acre of grass, assuming all of the area under grass or oaks is grazable. Following the "take half, leave half" rule, you could graze 537 lbs/acre of that. An animal unit requires almost 10,000 lbs/year, so you would need 18.6 acres per animal unit (10,000/537 = 18.6). Most range science experts say that about half of the half that is "taken" does not really go into the livestock because it is trampled on, pooped on, coated with mud, contaminated with undesirable/toxic weeds, or eaten by rabbits and insects, so that only 25 percent is actually available. This means that of the 1,075 lbs/acre of forage available, only 25 percent, or 269 lbs/acre, will wind up in the cow. Thus it will actually require about 37 acres to feed one animal unit.

Of course, in years of poor rainfall less grass is produced, and if the number of cattle is not reduced correspondingly the range will be even more overgrazed and the potential productivity will decline even more. Or, if some of the land is underutilized by the cattle, then the grazable acres will be less and the number of animal units that can be accommodated will also be less.

You may not know for sure what the number is, and it will vary from year to year with rainfall and past years' grazing pressure, but there is definitely an amount of forage that the land will produce that can be harvested without reducing the ability of the land to continue to produce that amount of forage, that is, without degrading the habitat. An animal or a collection of animals requires a certain amount of forage on a daily basis. Clearly, the goal is to keep the number of animals below the number that require more forage than can be harvested from the land without taking so much of the grass that the plants are damaged and the desirable forage plants begin to decrease in size and number. Put in more familiar terms, the stocking rate should never exceed the carrying capacity.

A hypothetical example should help clear up any confusion: If the carrying capacity of a property is 25 grazable acres/AU, then a stocking rate of 30 grazable acres/AU is *less* than the carrying capacity and one of 20 grazable acres/AU is *more*. If this property consists of 100 grazable acres, then its carrying capacity is four animal units. If you stocked it at a rate of 20 grazable acres/AU, you would have five animal units on the property, which would then be overgrazed.

In practice, since one never knows how much rain will come in a given year, prudent ranchers stock somewhat below what they think the carrying capacity is so they are not overstocked if rainfall is below normal. If a real drought begins and the pasture is showing signs of overuse, some animals should be sold or penned and fed to keep them off the pasture. Another reason to keep stocking rates below the theoretical carrying capacity is that, whereas animals need to eat every day, grass grows faster in the spring and early fall and maybe not at all in the winter. So there may be enough grass produced over the course of the year but not enough at all times of the year.

It was common for early settlers to believe that heavy rain years were the norm and droughts were an aberration, even though both probably occurred with equal frequency. Even throughout much of the twentieth century, many ranchers continued to maintain that belief. Many ranchers also thought that, if they still had grass at Christmas, then they didn't have enough cows. Both views caused a great deal of destruction of Texas ranges.

Animals always spend more time in the general vicinity of a water source than out on a dry hillside at the far corner of the pasture. This means that the area nearer the water may become overgrazed even though good grass is left untouched in other areas. This lack of uniform grazing becomes greater as the distance from the water increases, if the hillside gets steeper or rockier, or if the path to some parts of the pasture goes through dense cedar thickets. It is tempting to think that, if there is good grass somewhere in the pasture, the animals can go there if they want to, and if they don't it must be because they aren't that hungry. That may seem reasonable to us, but a cow doesn't always see the world as we do—maybe because we aren't the ones who have to haul our 1,200-lb bodies up that rocky hill in the heat just to find grass a few inches taller.

For all of these reasons, and more, to avoid overstocking a rancher has to be conservative in his stocking rate, even though that may mean fewer animals to sell at the end of the year and less income. And this approach may well improve the condition of the pasture enough that in the long run the productivity of the land is greater and the carrying capacity higher accordingly. The analogy of spending all of your income or putting some in savings for next year and later is apt. So is the idea of "not eating your seed corn."

From the preceding discussions, it should be easy to understand how many properties have been overgrazed, some almost continuously, over many years. Ranching in Texas is a tough business, and few landowners actually make significant income from cattle alone these days. Hunting leases and other sources of income frequently make the difference between being able to hold onto the land and having to sell out.

It is also true, though, that many enlightened ranchers have used the best management practices to improve or maintain their land in excellent condition, and they serve as role models for all of us. Their success proves the validity of many of the currently accepted management practices for Hill Country ranching that are discussed in this book.

Most relatively new landowners in the Hill Country are not trying to make a living by raising animals, however. Some folks raise animals just because they like them and enjoy having them around; some keep animals only to qualify for agricultural tax valuations. In fact, most new landowners don't own enough acreage to raise enough animals to make any significant income even under the best of conditions; they just moved to the country to enjoy the rural lifestyle. And even though income may not be the main purpose for having the animals, most folks have at least a general desire for their property to be in "good shape," to be "healthy," and to "leave it better than they found it." To do that, one has to be able to assess the condition of the land and understand what is good and bad about that condition, and to know what can be done to either maintain it in good condition or to improve it if necessary. This is what we discuss next.

How to Recognize Overgrazed Land

Whether you have lived all your life on your place or just bought it, you need to be able to study it critically and make a reasoned, scientific judgment as to its condition. This can be hard to do, kind of like trying to judge how smart and good looking your kids are; you love them, so you may not be impartial. But there are signs to look for and ways to measure that will help you make this assessment. Overgrazed land has less grass, less "good" grass, more bare ground, more inedible weeds, and more erosion than well-managed land. I guess all of that is obvious, but the trick is to be able to tell whether your place fits that description.

Having less grass means having less total forage for the animals, which obviously means fewer total pounds of forage per acre. You can measure this by clipping all of the grass in a specified area, say one square yard, drying it, weighing it, and calculating the amount per acre. A single measurement by itself doesn't mean that much, but doing the same measurement at the same time every year and comparing the numbers may be useful. You can also take several measurements from different parts of your property, possibly including some areas where the animals are excluded.

Less accurate but far faster is simply to measure grass height (height of grass blades, not seed heads) in several places and at different times of the year and record all of your measurements/estimates. You can also make small fenced exclosures and compare the grass in those to the amount outside being grazed. Finally, you can compare your grass amounts to that of your neighbors, but that is useful only if you know how many animals your neighbors have relative to what you have.

All of these things give you some numbers to compare, but they help only if you keep good records and compare them alongside your records of numbers of animals and rainfall. It is difficult to put a number on how much grass you "should" have, since this depends on so many variables. In general, though, pastures in the central part of the Hill Country that most people would consider to be in good condition probably have 2,000–3000 lbs/acre near the end of the growing season (October). Pastures that are clearly overgrazed may have a tenth of that. Then again, places with very thin, rocky soil may have very little grass even with little or no grazing.

By less "good" grass we mean less of the really nutritious, palatable grasses and more of the less nutritious, nonpalatable grasses. Good grasses that you like to see in a pasture are any of the "Big Four" tall prairie grasses: big bluestem, little bluestem, switchgrass, and yellow indiangrass. Other grasses that are good to find are sideoats grama, buffalograss, Canada wildrye, plains lovegrass, silver bluestem, cane bluestem, southwest bristlegrass, and Texas wintergrass. Grasses you do not like to see are hairy grama, hairy tridens, Japanese brome, purple threeawn or any of the other threeawns, Texas grama, tumblegrass, and windmillgrass. Well-managed land has more of the first group and less of the second, and overgrazed land just the opposite. In fact, a predominance of the poorer grasses is probably a better measure of past overgrazing than the amount of grass present at any one time.

Bare ground is, of course, an indication of less than optimal grass cover, but it is also largely responsible for an excess of weeds and contributes significantly to erosion. Well-managed properties have little bare ground. One of the best ways to measure and record the amount of bare ground is to do a transect study, whereby you step off a given distance between two points, and as you walk you record at every step whether your right foot touches bare ground, grass, forbs, or litter (dried leaves). Doing this at the same time each year helps you find out if things are getting better or worse.

Although well-managed pastures certainly have a mixture of forbs as well as grasses, what you do not find is a preponderance of inedible weeds, such as Mexican hat, mealy blue sage, ragweed, queen's delight, thistles, buffalo bur, or snow-on-the-mountain. Heavy infestations of any of these may indicate past overgrazing that has reduced the grass cover, thus giving rise to bare ground and the subsequent increase in forbs.

Everything we have discussed about overgrazing so far assumes that the animals in question are cattle. This is not because cows are the only animals that can overgraze a pasture. It is simply because cattle are the most familiar to most people, and they make for an easier discussion of the general principles of grazing.

If you have other animals in addition to, or instead of cattle, the most important things you need to know are what they eat and how much they eat. Horses and their kin (donkeys, mules, zebras) eat almost exclusively grass. Sheep eat more forbs and browse than cows, but they still prefer grass. Goats and most exotic ungulates (goats in deer clothing) freely move from browse to forbs to grass and back depending on what is good today.

The complication here is that most of these other animals don't eat just grass but a combination of all three types of food, and it is considerably harder to judge how much browse and forbs are available and how much is being eaten. Thus, assessment of overgrazing/overbrowsing the land is more difficult, although the principles remain the same as discussed for cattle. One final complication is that white-tailed deer are almost exclusively browsers and forb eaters, so they are in competition with the sheep, goats, and exotics, and vice versa. We examine overbrowsing in chapter 8, after which it will be easier to assess the status of your property in terms of the effects of all the animals.

The Consequences of Overgrazing

Overgrazing leads to a reduction in the size of the grass plants, which reduces the amount of new grass production in the next growing cycle. In addition, over time grasses that are less palatable replace the better, taller grasses, so that even in times when overgrazing does not occur the productivity of the land is reduced.

In addition to less grass being produced and fewer animals being raised, overgrazing has several other consequences. One is that the ecology of the area is changed. The replacement of "good" taller grasses with "poorer" shorter grasses makes for not only a less productive rangeland but a poorer habitat for turkey, quail, and other wildlife. Along with shorter grasses in an overgrazed pasture is increased bare ground. Bare ground is usually required for forb reproduction, so one might expect more forbs, and thus more food for forb eaters and seed eaters such as turkey and quail. Unfortunately, along with overgrazing we have over-abundant deer, which tend to eat all of the edible forbs, leaving only the less desirable inedible forbs. Some of these are still good seed producers, such as croton (dove weed), but the overall net result is not just shorter grass and more forbs but more forbs that are inedible—some of them (thistles, nettles, grassburs) really problematic.

The increase in bare ground has other consequences. Bare ground gets hotter than soil under grass, by as much as 20 degrees. This not only dries out the soil faster, it reduces the numbers of microorganisms and small animals in the soil, thus decreasing the porosity and fertility of the soil. Bare ground can also develop a hard crust—an impermeable layer on top. All of this means that water soaks into bare ground slower than it does under grass, so that less rainfall soaks into the soil and more runs off. Bare soil is susceptible to having small bits of soil dislodged by raindrops (which hit the ground at 20 miles/hr), thus increasing the amount of soil carried off even more.

More runoff and less rainfall soaking into the ground result in less water infiltrating deep underground to recharge aquifers, springs, and seeps. Less spring flow means less water available to us all. So the way the land is managed really does affect us all.

What to Do about Overgrazing

Although the details of what overgrazing does to the land, the many factors that enter into understanding it, and the consequences of overgrazing are somewhat complicated, the solutions are not. The solution to overgrazing is to *rest the pasture.* There is no secret here at all. There are no long, complicated procedures or actions. It requires no new equipment or extra manpower. And it doesn't even require any extra expenditure of money. Of course, it may limit income for a time. Simply removing the animals almost always allows the land to recover, at least partially. It seems to me that too few people take advantage of the provision in the agricultural tax valuation requirements that allows you to remove animals for two of every seven years and still maintain the valuation.

How complete and how long the pasture needs to be rested depends on several factors: how bad a shape the land is in, how much rainfall you get, how deep the soil is, how long the land has been overgrazed (thus how good the native seed bank is), and perhaps how much control you have over exotics and white-tailed deer. Resting the pasture usually means removing the livestock from the area for at least one growing season, maybe more. Less extreme measures may help, but they probably won't make the kind of improvement you are hoping for. Some actions may be taken to increase the rate of recovery, such as prescribed burns, seeding, or terracing, but these are not usually necessary to improve a pasture greatly. Mother Nature is a great healer, if we will just let her be.

Rotational grazing is beyond the scope of this handbook. The concept is simple enough, but putting it into practice successfully is more complicated than one might think. In the ideal case, where land, fence, water, and other resources are unlimited, one would divide a large pasture into twenty or so equal smaller pastures, put the entire herd into one pasture for two to four days, and then rotate them to the next pasture, giving the grazed pasture forty to eighty days of rest. This practice mimics the action of bison herds when they grazed an area very hard for a short period of time and then moved on, allowing a long period of rest for that land.

Few people can accomplish this ideal case, sometimes referred to as a short duration system. But even a two-pasture system is beneficial, and a three-pasture, one-herd system is common and can be quite effective in improving the quality of all pastures. Figuring out when to graze a given pasture and when to move the herd is complicated, but the decisions are based on the condition of the pasture being grazed, how long the other pastures have been rested, and what time of year it is. AgriLife Extension and NRCS agents can help landowners develop plans for their property, but even the best plan must sometimes be altered depending on rainfall, past condition of the pasture, and stocking rates.

This is a good place to discuss different types of animals and how they are managed as this relates to the grazing pressure on the land. Many folks who want to have livestock choose cattle, and most of those choose a cow/calf operation. This is probably the most difficult livestock to keep in terms of requirements of animal husbandry. First, this type of operation requires you to own or lease a bull, or to pay for artificial insemination, all expensive options. Then there can be problems with cows giving birth, especially young cows, and then the doctoring, tagging, possible castrating, and dehorning the calves, all of which require manpower, know-how, and special equipment. Finally, it seems that once a landowner has cows for a while, he becomes attached to them, which makes it difficult to sell some of them when range conditions deteriorate. So people wind up keeping cows, paying for supplemental feed, and overgrazing the land.

A much simpler, easier way to have cattle is to buy stocker calves (weaned calves usually 400–500 lbs) and grow them out for a few months and then sell them. This allows for many fewer animal husbandry issues and makes it much easier to adjust the animal numbers to the amount of forage you have. Another approach is to lease your property to a cattle owner with terms that allow you to dictate the number of animals and the timing for introducing and removing them. That way the cattle owner is responsible for the care of the cows and calves, you have very little work to do, but you can still control the amount of grazing that occurs on your property.

Many people choose goats because they are smaller and easier to handle. This has its advantages, but also disadvantages: goats require much better fences and gates, they can be destructive to your vehicles and structures, and they are more aggressive at searching for food and can thus overbrowse an area if not closely monitored and their numbers strictly controlled.

Raising sheep is another alternative. They are not quite as hard to contain or as destructive as goats. Wool sheep present the problem that it is difficult to find anyone to shear them if you have only a few, and it may not be a job you want to learn. So hair sheep may be a better alternative.

Sheep and goats have predator problems—coyotes, but also wild dogs and maybe even the neighbor's pets.

Having raised sheep and calves as a kid in 4-H, I can well understand the attraction of raising these animals, but I also remember the drudgery of getting up before dawn on cold winter mornings to feed and water them before going to school. It is a big responsibility, and it is hard to take a day off.

There is an understandable tendency when trying to improve a pasture to want to scatter grass seed around, especially on bare areas, in the belief that, if what is missing is grass, adding grass seed will fix the problem. A more enlightened approach is to ask why no grass is growing in this spot or why the grass is so thin and scattered. In general, adding grass seed is not successful unless the reason for no grass is that there is no seed in the ground, which is hardly ever the case except in areas where topsoil has been scraped away. In general, one should ask these questions: What current management practice is responsible for the current condition? Can the condition change without a change in the management practice? There are times and places where reseeding is helpful, but, as Bob Lyons likes to say, "It is the most risky thing you can do in land management," meaning that it has the lowest probability of success. It should be noted that Lyons works mainly in the western part of the Hill Country, which is dryer and hotter than most of the Hill Country. The probability of success with reseeding increases with increased rainfall.

Most bare ground begins to grow grass with sufficient rainfall and lack of grazing. Cases where this is not the case can be very thin soil over rock, so that the soil cannot support denser grass; constant trampling/grazing by animals in high-traffic areas; and soil contaminated by fuel spills or chemicals. Grass plants produce a lot of seeds, and they get dispersed very well, so if you have grass in the area but none in certain bare areas, it is probably not because there is no seed in the ground.

It is certainly true that, if you are missing some of the better grasses (big bluestem, little bluestem, switchgrass, indiangrass, sideoats grama) or some of the better perennial forbs in areas where they should be growing, seeding those areas with these grasses and forbs may well improve the diversity of vegetation and the quality of the pasture. But subsequent management will dictate whether the improved diversity is sustainable or reverts to the previous condition.

8

Overbrowsing

For our purposes I define browsing as animals eating the leaves of woody plants or forbs, to be distinguished from grazing, which is eating grass. Herbivores tend to be grouped into three classes: those that eat primarily grass, those that eat primarily browse (leaves of woody plants) and forbs, and those that eat comparable amounts of all three. Virtually all animals that eat browse also eat forbs, and vice versa, but not all animals that eat those two types of forage can eat grass. Browse and forbs both contain more protein than grass. Grass is the most lignified and thus the most difficult to digest, as discussed in chapter 7.

Woody plants include trees, shrubs, and vines that are perennial and grow on previous year's wood. Forbs are herbaceous, broad-leaf plants that can be either annual (die back to the ground in winter; new growth is from seeds) or perennial (sprout new growth from the root base the next year). The term *broad-leaf* is used to distinguish forbs from grasses. Most people think of forbs as weeds and wildflowers.

Shin oak is not a favorite food for deer, but they readily eat it wherever their more favorite foods are unavailable. Note the three-year-old root sprouts inside the exclosure and the total lack of any such vegetation outside.

Food Selectivity

Cattle and horses tend to be almost exclusively grazers, but they occasionally pull down a tree branch to eat the leaves, especially in winter, or take some types of wildflowers when they are present. White-tailed deer are almost exclusively browsers, but they occasionally eat some grass, especially young shoots, or if they are quite hungry. Sheep, goats, and most exotic ungulates eat all three types of food, depending somewhat on what is available and somewhat on just what they happen to like best. Of the common exotics, blackbuck antelope and oryx tend to eat more grass and less browse and forbs than most of the other exotics, whereas axis deer tend to eat browse and forbs and less grass than the others.

White-tailed deer are at a disadvantage in competition with exotics because their rumen cannot digest enough grass to live on, so if grass is all there is available, the white-tail suffer. More generally, animals that can live on three types of food outcompete animals that can live on only two types. In an experiment at the Kerr Wildlife Management Area, six white-tailed deer and six sika deer were placed in a 96-acre high-fenced pasture. Initially their numbers increased to sixteen and fifteen, respectively. But then the animals had eaten all the browse below the browse line as well as all the forbs, so grass was all that was left. After nine years, there were sixty-six sika and no white-tail in the pasture.

From a land manager's standpoint, it is important to understand what every kind of animal on your place eats, as well as how much, in order to assess the impact of each species on the land. To the extent that different species eat the same type of food, they are in competition, but if they eat different types of food, they do not compete. Strict browsers do not do well in an open grassland and certainly not on an irrigated coastal bermudagrass field. Strict grazers do not do well in a dense woodland with little grass.

The discussion in chapter 7 about grazing is oversimplified in assuming that the only plants eaten by animals are grass. But we know that cows can eat an average of about 10–15 percent browse. The principles of carrying capacity, stocking rates, animal units, "take half, leave half," and grazable acres still apply to any single species eating more than one kind of food, or for that matter to several different species eating different foods. The details are a little different between grazing and browsing, the different plant types' growth patterns and response to being eaten may be different, but the ideas are still the same. Calculating or measuring how much of the different plant types you have or how much of each is being eaten is harder than for grass alone, but the principles are still the same.

If the stocking rate of all animals exceeds the total carrying capacity (to produce grass, forbs, and browse), then the habitat will be degraded. Indeed, if the stocking rate of any one species of animal exceeds the carrying capacity for the land to produce the type of food required by that animal, then the habitat is degraded. Even if the animal can then switch to a different food type, it has still overgrazed/browsed that one food source.

The Consequences of Overbrowsing

For the purposes of discussion in this chapter, we consider browsing simply as eating forbs and woody plants, and overbrowsing as taking so much of these types of plants that the total biomass as well as the species distribution are significantly altered. Just as overgrazing alters the species distribution by weakening the favored grass species and thus allowing the less-favored species to increase, a similar process occurs with forbs and woodies.

Forbs can be either annuals or perennials. Annuals live, produce seed, and die in a single year, and next year's plants all come up from seed. Browsing not only weakens plants but can prevent them from making any mature seed. Obviously, if this happens repeatedly to annuals, eventually there will be little if any seed left and the species will disappear from that area. If perennials are browsed, their seed-producing ability may be limited or eliminated for that year, but being perennial they can come back next year from the roots. Such plants will be weakened, however, for the same reasons as discussed earlier for decreaser grasses, and though they may come back the next year they are likely to be weaker. Eventually, if these plants are repeatedly browsed, they too will disappear from the area.

The simple result of overbrowsing of forbs is selection for the less favored, less edible ones through elimination of the more favored, most palatable ones. We as Texans take great pride in our famous Hill Country annuals—bluebonnets, Indian blankets, and

Indian paintbrush—as well as perennials—Engelmann daisy, Maximilian sunflower, and pink evening primrose. In reality, however, there is a lot more Mexican hat, prairie verbena, mealy blue sage, queen's delight, snow-on-the-mountain, doveweed, cowpen daisy, and frostweed in Hill Country pastures because those species are a lot less favored by deer and other browsers than the former group. One of the reasons the TXDOT program of seeding the roadsides with bluebonnets and the others is successful is that the deer population is lower along highway rights-of-way than in adjacent pastures. So, just as overgrazed pastures can be detected by observing the species distribution of the grasses, forb species distribution can reveal overbrowsed pastures. I have observed cowpen daisy (and not much else) growing in abundance in small deer pens at the Kerr Wildlife Management Area research facilities where there were dozens of deer in pens of less than one acre. Similarly, on a clearly overstocked goat ranch I observed acres of beautiful mealy blue sage, but not much else.

The result of overbrowsing of woody plants is pretty much analogous to overgrazing grass or overconsuming forbs. The total amount of available vegetative material is reduced and the species distribution is altered, again in favor of the least favored species at the expense of the most favored. The visual effect of overbrowsing of woody plants may be greater than that of overgrazing because overbrowsing of woodies robs us of a whole layer of vegetation. Overbrowsing of trees and shrubs has created a **browse line**. On most Hill Country properties, there is essentially no woody plant (other than cedar) with leaves nearer than 5 feet off the ground, the height most white-tailed deer can reach by raising up on their hind legs.

If you ask average landowners to think about their vision of a typical Hill Country tree and then to draw one, they inevitably draw a tree with a space between the ground and the first limbs and leaves—the lollypop shape. If asked if this is the way native trees grow naturally, most everyone will say yes. And they will be wrong. It is true that, as trees grow, their lower limbs become shaded by the upper limbs and, getting less sunlight, produce fewer leaves and eventually die and fall off (we are talking decades here). But what is not true is that there are no leaves below some certain height, say, five feet. Native Hill Country trees naturally have perimeter limbs that extend to or almost to the ground wherever they are not so crowded by neighboring trees that they can't get sunlight. Or at least that is the way they would grow if it were not for browsers eating all of the lower leaves.

It is not the loss of the bottom five feet of leaves from a 30-foot live oak tree that is the problem, however. It is the lack of any new trees, growing either from seeds or acorns or sprouting from the roots of the mature trees. Where are the baby trees, where are the teenagers, where are the trees that will replace our large mature hardwoods when they die? The answer is that there are none, save a few growing over a creek bank or inside a large cactus, inaccessible to the deer. Deer are changing the landscape of the Hill Country as well as the ecology.

Birds that like to build their nests three feet off the ground, like the black-capped vireo, are having a hard time finding nest sites. Small animals that like to den or nest under protective shrubs are likewise out of luck, many being now more vulnerable to predators. Many species of small trees, shrubs, and vines that provide food for all manner of native fauna are becoming scarce. The deer are damaging not only their own habitat but that of many other critters as well. We have probably seen the highest population of woody species (except for cedar) in the Hill Country, and will begin to see a decline in both numbers and number of species as time goes by.

One might ask, How do we know the fault is with the deer and not some change in the climate or environment? The answer is that in places where the deer numbers have been controlled, such as the Kerr Wildlife Management Area (one deer per eight to twelve acres), there are many woody plants growing below the browse line as well as mature trees with limbs and leaves down to the ground. In my own experience, we high-fenced an acre of overgrazed, overbrowsed oak savanna, and within three or four years we saw black-jack oak and post oak limbs literally reaching the ground and live oak sprouts three feet tall. In addition, escarpment black cherry, hackberry, and prairie flame-leaf sumac began coming up all over the area, even though none of these three grows within several hundred yards of our fenced area. The oaks outside the fenced area continue to show the usual browse line, and the only sprouts of any kind to be found are those inside small exclosures protected from the deer.

If the Hill Country once had fewer trees than it does now, as described in chapters 3 and 4, what would be so wrong with return-ing to that type of ecosystem? The answer is that, although there were once fewer trees on the open prairies than we have now, there were hardwoods and shrubs growing on the slopes that provided the food sources and habitat that are now missing. In other words, overbrowsing is occurring not only on the flatter, more open areas but also on the slopes and along the creek beds. Another reason this reduction in woody species is a problem is that it is happening so quickly that many creatures are having trouble coping with the change; they can't adapt fast enough. There is also a certain hu-man, aesthetic component to this: we all think of the Hill Country as having many trees, and the thought of our land becoming a tree-less grassland is upsetting to many people.

How to Recognize and Assess Overbrowsing

Earlier we discussed assessing overgrazing by studying the total amount of grass in a given pasture as well as the particular spe-cies present. The presence of good, desirable grasses in abundance indicates a healthy range, whereas the predominance of small, less palatable grasses indicates overgrazing. Similarly, the presence of significant amounts of edible, desirable forbs indicates only light consumption, whereas a predominance of inedible forbs, even if in abundance, probably indicates overconsumption.

White-tailed deer eat almost any woody plant in the Hill Country, but they have clear preferences, just as all other animals do. These preferences differ somewhat from individual to individual, and from location to location. Also, if none of their favorite food is available, then whatever is no. 2 becomes their new favorite food. (Similarly, if we go to a restaurant and it is out of steaks, we will eat hamburgers, or chicken, or pork chops. But if there is steak, we will probably pass up the bologna sandwich, and if a deer can find Spanish oak and kidneywood, it is probably not going to touch the cedar.) So knowing the deer food preferences and observing what is and is not being eaten can tell us a lot about the extent of overbrowsing and thus the deer population, relative to the amount of food available. One such food preference list published by the Texas Parks and Wildlife Department appears in this book as appendix 1, where various native plants (woody plants and forbs) are listed in categories of preferred deer food, good deer food, low-quality deer food, and little utilized species.

If a survey of the woody plant leaves found below the browse line and within reach of the deer reveals any of the preferred deer food or a significant amount of the good deer food, then the browser population is probably not so great that overbrowsing is occurring. Obviously, if no browse line is observed, then the browser population can be considered well controlled. If, however, only very little of the second group is found, then overbrowsing is occurring to some extent. And if there is evidence of significant browsing on the two lesser-used group of plants, then it is likely that severe overbrowsing is occurring, possibly indicating a deer population of one deer to every two or three acres. Similarly, if any plants from the little used forb category are eaten, this probably represents an overpopulation of forb eaters. Of course, it is important to keep in mind that white-tailed deer are not the only browsers in the Hill Country; most exotics, as well as goats and to some extent sheep, can contribute to an overbrowsed condition.

What to Do about Overbrowsing

The first thing to do about overbrowsing is remove or reduce the numbers of the offending animals, but it is important to be able to identify which animals are responsible. If you have ten pigmy goats and a hundred or so white-tailed deer, removing the goats probably won't fix the problem.

This is as good a time as any to state that goats' bad reputation is only partially deserved. Any animal can abuse the land if the landowner allows it to. It is the landowner who is responsible for controlling animal numbers and therefore their impact on the land. It is probably true that goats are, pound for pound, harder on land than most other animals, because they more aggressively search out the last edible morsel and probably have a longer list of species they will eat than most other animals. We have all seen ranches almost totally denuded of vegetation below about five feet. These "goated out" places are not, however, inevitable, and I have seen many ranches with goats where one would not suspect the presence of goats from the condition of the vegetation.

Some animals (feral hogs, axis deer, white-tailed deer) may be more difficult to control than others (livestock), but landowners who take the attitude that "there is nothing I can do" are contributing more to the degradation of the land than the animals. At least with livestock, any willing landowner can reduce their numbers or remove them entirely.

When we discussed overgrazing, one of the remedies suggested was rotational grazing, which rests the pasture for a few months every year. This rest period, if occurring during the growing season, allows the grass to recover from being eaten. Unfortunately, this same practice does little to improve an overbrowsed landscape, because of the difference in time scale and the growth characteristics between grass and woody plants. Indeed, resting does allow an overbrowsed landscape to recover, but it may take years to do so.

A perennial grass plant can grow from a close-cropped dormant state in January to a full-size, reproducing plant by September, with most of that growth occurring between April and July. In contrast, it takes a post oak sprout ten years or more to become mature enough to survive being browsed (i.e., when it is at least ten feet tall with hard, mature bark)—so a temporary rest is not going to save it. What will save that post oak is a small enough population of browsers that they miss the sprout until it has several leaves. Then, when a browser does come by, it takes only some of the leaves, and next year the sprout manages to produce even more leaves and even more escape browsing, until it becomes a big sapling. Most of its brothers and sisters may be eaten away, but enough will escape the predation and grow to maturity to become the replacements for the parent trees.

What population of browsers is low enough to allow that? The biologists at the Kerr Wildlife Management Area have pretty convincingly shown that a white-tailed deer to every ten to twelve acres on well-managed land is a "safe" population in terms of allowing the habitat to regenerate itself. In other words, one deer to every ten to twelve acres is the carrying capacity of the management area. And that assumes no other browsers such as exotics or goats, but it does allow for cattle. There are some areas in the Hill Country that can carry as many as one deer to every six to eight acres with little detrimental effect. Unfortunately, bringing the deer population of the average Hill Country ranch down to even that level, especially if it is a small area and is low-fenced, is quite difficult. We discuss some of the issues involved in reducing white-tailed deer populations and maintaining them at these low levels in chapter 11.

9

Cedar Management

DESPITE rumors to the contrary, Ashe juniper, which we call cedar, is native to the Hill Country, and though historical reports indicate that juniper woodlands existed over much of the Hill Country, there was overall less of it at times in the past. It is, in fact, almost the perfect plant for this area, for it has very few diseases or pests, is quite drought tolerant, can grow in almost any soil (and very little of it at that), and reproduces well. Unfortunately, for all of the same reasons, it is also one of the biggest pests. Cedar does provide cover for many different animal species and food for some as well. As with most pests, it is not so much the individual properties that are a problem, but the numbers.

Why Manage Cedar?

As discussed in chapters 3 and 4, cedar has taken over much of the Hill Country. Beginning in the mid-1800s when Europeans began settling this area, they began to overgraze the grasslands and to suppress fire, both of which enabled cedar to become established in areas where it had not been previously. The problem is exacerbated because nothing eats cedar, or at least not enough to affect its growth significantly. White-tailed deer and goats eat some cedar, especially in winter and when their preferred food is scarce, but even in areas with high white-tailed deer populations cedar continues to increase in size and numbers. Therefore, we find excess cedar just about any place where cedar has not been controlled by landowners.

The Kerr Wildlife Management Area high-fenced a 96-acre section of savanna covered with grass and oaks forty years ago and left it to "let nature take its course." There were no animals inside the exclosure, and no human activity took place there. The area is now a complete cedar brake—a large area with dense stands of large cedar bushes nearly touching each other—with no grass and mostly rocks covering the ground where the soil has been lost. Such a monoculture provides habitat for very few species and is quite undesirable.

Cedar has several properties that are detrimental to surrounding vegetation. It has the ability to outcompete most other native plant species, thus forming the monoculture mentioned above. It does this by reproducing quite prolifically in the normal fire-free environment of most of the Hill Country and by being able to crowd out underlying vegetation by blocking sunlight with its dense, evergreen canopy and lower limbs. In addition, it appears to outcompete most neighboring vegetation for soil moisture, so that in thick cedar brakes, even in space between the trees, one finds little or no native grasses, forbs, or woody species. Finally, there is evidence that cedar trees intercept some of the rainfall and allow it to reevaporate back into the air so that the water never reaches the ground, thus reducing the effective rainfall under the cedar.

Cedar and Water

This topic is much talked about and fraught with misinformation. But because it is a subject that many people don't fully understand, it seems appropriate to spend some space here to discuss what we know and what we don't know. This topic very much reminds me of a favorite poem a colleague of mine used to quote, paraphrasing Mark Twain:

> *It ain't the things we don't know*
> *That pains us so*
> *But the things we know*
> *That just ain't so*

 I can tell you now that at the end there is not a definitive answer or clear guidance about cedar and water. But I can also tell you that in the final analysis, for a landowner trying to decide how to manage land, it really doesn't matter that much.

There seems to be no dispute that when it rains some of the water that falls from the sky wets the leaves of vegetation and stays there until the rain is over and the water on the leaves reevaporates into the air. The result is that a smaller amount of water, measured in inches of rain, reaches the ground under vegetation than it does on bare ground. The question is, Is the amount of water that is intercepted by leaves different for various types of vegetation, specifically native grass, hardwood trees, and cedar? Some research seems to indicate that in fact, in light rains, cedar does intercept more water than grass, and that hardwood trees may be intermediate. The exact numbers depend on many variables to be discussed below, but as the intensity of rainfall increases, any difference in interception between the different vegetation classes diminishes.

There are many instances throughout the Hill Country where landowners have cleared cedar from their land and subsequently observed springs or seeps producing water where none had been seen before or producing more water than they had before. The Bamberger Ranch Preserve in Blanco County is the most famous example of this, but there are many others. These results lead to the logical conclusion that, if cedar were removed, more of the rainfall would reach the ground, soak in, and feed springs and seeps. It is also true that the results can be explained in a different way—

A common sight within a dense cedar brake: large cedars growing close together; surface soil eroded away, leaving mainly stones; and little if any grass or forbs growing between the bushes.

by cedar "using" more water by the process of evapotranspiration than other vegetation, so that its removal results in less water being taken from the soil, and thus more water remaining to recharge an aquifer.

The best that can be said at this point is that, if you were to remove cedar from your land, you *might* see an increase in spring or seep flow and you *might* see an improvement in the growth of other vegetation, especially grass and forbs. Then again, you might not. If you already have a spring or seep on your property that flows at least a little during wet periods, the probability of enhanced flow after cedar removal is increased. Why is the effect of cedar on the fate of rainfall, also called the water cycle, so uncertain? Because there are so many variables. Here are some of them.

- Rainfall
  - Total amount
  - Size of rainfall events
- Soils/Geology
  - Type of Soil
  - Depth of soil
  - Type of underlying rock
    - Fractured limestone
    - Solid limestone
    - Granite
  - Porosity of rock
  - Permeability of rock
  - Slope
- Cedar Trees
  - Size of individual trees
  - Density of trees
  - Amount of litter beneath the trees
  - Porosity of soil underneath
- Replacement Vegetation
  - Amount and size
  - Grass
  - Forbs
  - Woody plants
  - Root structure
  - Amount of litter
  - Soil permeability

The **water cycle** describes all the different things that can happen to rainwater. For example, rainwater falling from the clouds can either evaporate in the air, be caught by vegetation and evaporate back into the air, or fall to the ground. Water that hits the ground can evaporate, soak into the ground, or run off. Water that runs off can evaporate at some stage, become part of a stream, river, or lake, and either be used by humans, nourish plants along riparian areas, or flow into a larger body of water such as the Gulf of Mexico. Water that soaks into the ground can be retained by the soil or it can seep deeper underground, where it can become part of an aquifer or flow laterally to feed a spring. Water that stays in the soil can either evaporate directly into the air or be taken up by plants and become part of the plant structure or be evaporated.

Here the discussion becomes partly ethical and partly political. What do you want to happen to that raindrop? The answer may depend on who you are. If you are the landowner, you want as much of that water as possible to soak into the ground, either to replenish soil moisture and thus nourish your vegetation or to recharge aquifers that feed springs or seeps so you can maintain a riparian area or provide water for livestock or wildlife. On the other hand, if you are a city dweller, you may want all of the water that falls on the land to run off and fill up the lake that supplies your city with water. Or, in the case of an aquifer that is directly recharged by rainwater, such as the Edwards Aquifer, you may want all of the water to run off and recharge that aquifer. But too much runoff means flooding, sediment buildup in lakes, and much of the water flowing into the Gulf. It probably also means erosion on the land, and eventually that process will degrade the landscape. A more enlightened city dweller may realize that, if the water were to soak into the land and gradually issue forth in springs that feed rivers over a longer period of time, a more constant flow of purer water requiring less purification would result.

So, if we assume that the best thing to happen to that raindrop is to soak into the ground, what does that mean in terms of how the land should be managed? If one accepts, at least under certain circumstances, that cedar does intercept more rainfall than grass and may soak up more water from the ground than other vegetation, and that removing cedar may give rise to greater spring flow, then it is hard not to reach the reasonable conclusion that removal of large amounts of cedar from the watersheds of our major rivers would result in greater flow in those rivers. There are, however, no direct data to show that this is in fact the case, and there is even some historical information that indicates no correlation between the amount of cedar and river flow. So we are left with the question, Would removal of large amounts of cedar from the Hill Country give rise to increased river flow or aquifer recharge? Looked at differently, has the river flow decreased as more and more cedar has grown up in the Hill Country? Bradford Wilcox of Texas A&M, a leading researcher in this area, sums up the situation in a recent publication like this: "Fundamentally, I am posing the question that is puzzling me: has stream flow changed as a result of rangeland degradation, and if so how? Our current hydrological understanding would suggest that it should have, but so far the signal is not particularly obvious. Is this because there is in fact little change, or because we have not looked hard enough?"

The reason this question is so important is that some people advocate large-scale cedar removal as a way to increase water available to Hill Country, and especially San Antonio, residents. In fact, government programs have already spent millions of dollars to aid landowners in clearing cedar to accomplish that very purpose. If the water supply for all the Hill Country depends on how private landowners manage their land, do we as landowners have a moral obligation to clear cedar? What if it doesn't really make a difference? I can't resolve the issue of how directly your land management practices affect the water supplies of downstream cities here. Maybe future researchers will be able to resolve this conundrum.

Choosing Cedar for Removal

Walk over most Hill Country properties and you will likely see few if any young hardwoods of any kind, but you will quite often see numerous young cedar trees. Most long-term landowners can attest to seeing cedar increase in coverage on their land over a period of years, even after complete clearing of mature cedars. And this happens in areas of good native grass stands as well as in overgrazed pastures. It seems clear that, just as in the tract at Kerr Wildlife Management Area discussed above, most anywhere we don't control cedar will eventually become a cedar brake.

I said earlier in this chapter that too much cedar results in too little diversity of vegetation, which is not desirable in a healthy ecosystem. So whether removal of cedar gives rise to more water soaking into the ground and subsequently greater stream flow, managing the amount of cedar you have is a good thing to do. Notice I said "managing," not eliminating or removing all cedar. Some amount of cedar is desirable and is probably healthy on most properties. How much may depend on the nature of the property, what else is growing there, and on the landowner's plan or vision for the property. (Benjamin Franklin was probably right: "moderation in all things.") But managing the amount of cedar is, like managing the number of grazing animals or browsers, part of what all landowners must do to have a healthy habitat. And how cedar is removed may be as important as how much is removed in many instances.

Everything we have discussed above has to do with land management. But there is one other issue that needs to be taken into account when discussing cedar: fire. Because cedar has fine scale-like or needle-like leaves with a high surface area and because it contains terpenes (hydrocarbon compounds), it can be very flammable when dry. The moisture content of the leaves depends on the recent wind, relative humidity, soil moisture, and temperature. When the moisture and humidity are low and the temperature is high, cedar can be extremely flammable, more so than most hardwood trees and shrubs. Cedar on a slope below an overhanging structure is particularly dangerous because the heat from the fire below dries out the fuel upslope and causes the fire to move with greater intensity and speed.

The bottom line is that fire safety may not be a reason to clear cedar from all of your property, but it certainly is a reason to remove it from near your house. We address this topic in more detail in chapter 16.

Before you remove any cedar, you need to think about a few other things: what will make the land healthier, which cedar to remove, which to leave, how to remove it, and what to replace it with? Remember that cedar has some clearly redeeming features:

- Cedar is an important food source for some birds and an important cover and shelter for deer and many small animals and birds. Some old-growth cedar in the right conditions is essential for the nesting of the golden-cheeked warbler (see chapter 18).
- In areas of significant slope and with little grass below cedar bushes, the cedar serves to intercept raindrops and partially protect the soil from erosion.
- If you have very few hardwoods, trimming up some cedar bushes into "tree shapes" gives the appearance of more trees.
- In some cases, especially where there is significant accumulation of leaf litter beneath the cedar trees, the cedar aids in the formation of new topsoil with high amounts of organic matter in places where grass cannot grow.
- Cedar trees frequently grow in fractured rock formations and aid the infiltration of rainwater.

One reason I am cautious about cedar removal is that I have encountered many folks who have done extensive cedar clearing and then set about trying to get something to grow in its place only to find that the ideal plant was cedar. Another reason is akin to the carpenter's rule: "Measure twice, cut once." Don't rush into a clearing program without first giving it a lot of thought. You can always remove a cedar bush in the future, but you can't get it back if its removal turned out to be a mistake. Finally, heed the physician's oath to "first, do no harm." A cedar brake on a steep slope may not be the best thing you could have growing there, but it is not the worst either.

The problem with removing cedar from a steep hillside is that it is difficult to remove cedar from anywhere mechanically without disturbing the soil. Even if the cedar is removed by hand, its removal exposes the bare ground underneath the bushes. Exposed bare soil on a slope is extremely vulnerable to serious erosion, and lots of soil can be lost in just a few thunderstorms. The cedar trees on the slope may or may not be the reason the soil is bare, but they serve to intercept and break the rainfall, thus protecting the underlying soil. Since the soil tends to be thin on steep slopes anyway, and therefore cannot support really heavy grass stands, having cedar there may be preferable to removing it and having the soil wash away and pollute creeks and streams. And cedar on slopes still serves as cover for wildlife.

From the standpoint of wildlife habitat, some cedar is definitely a good thing. For instance, deer prefer to never have to venture out more than 100 yards or so from cover. An irregularly shaped patchwork of cleared and uncleared sections provides the ideal general habitat for most wildlife, producing a lot of "edge" area between dense cover and open land.

General Brush Control

Before we discuss methods, it is useful to think about brush control in general, and some of the basic principles that apply. Just about anything can be considered unwanted "brush" if it is so prolific or dense that it crowds out other desirable species and becomes a monoculture. Monocultures are never good habitat. In addition to cedar, in various places on the Edwards Plateau and adjacent areas, mesquite, prickly pear, beebrush (whitebrush), huisache, redberry cedar, shin oak, Texas persimmon, and even Texas mountain laurel can be considered pests.

One fundamental difference between cedar (blueberry cedar, Ashe juniper) and all of these other species is that it is actually the easiest to kill. Cedar dies and does *not* come back from the roots if it is cut below the lowest green leaf. All these other plants sprout back from the roots or stumps, creating a much more difficult situation.

In theory, there are four types of brush control to consider: biological, chemical, mechanical, and fire.

Biological control. Biological control means having some animal eat the offending plant. The problem is that, almost by definition, if something is so prolific that it becomes a problem, it almost certainly is not eaten to any great extent by any common livestock or wildlife. Goats are often quoted as being able to "control" cedar, and under a very narrow set of conditions they may be able to. But that is only on land that has already been cleared of standard-size cedar, leaving only the problem of small cedar bushes sprouting from seed dropped by birds and other wildlife. Even then, goats eat a relatively small amount of cedar and still eat mostly other things first, since cedar is very far down on their list of favorite foods. Also, a goat can digest only a small amount of cedar per day. So goats are going to work only in places where the cedar size and quantity and the abundance of other forage are just right so that the animals eat the cedar before they take too much of more desirable forage—not a likely condition for most properties.

Chemical control. Chemical control essentially means herbicide treatment of the offending plants. For the most effective herbicide formulations, the Texas AgriLife Extension has a series of brochures called "Brush Busters" that describe the preferred methods for most of the common brush species in detail. Many people are hesitant to use chemicals on their land if it can be avoided, and rightly so, I believe. For some conditions of certain kinds of brush, however, chemical treatments may be the best choice. In those cases, I would try hard to use just the optimal amount of chemical as described by the brochures and to do so under weather conditions that require the least amount of the chemical. Most experts suggest chemical treatments of cedar only for plants under about three feet tall; anything larger takes too much chemical.

For other brush species, a method that uses the minimum amount of chemical is a "cut stump" treatment, which involves cutting the plant off just above the ground and immediately spraying the stump with a herbicide. This is usually as effective if not more so than other methods for root-sprouting plants.

Fire. For land that has been cleared of large cedar and is having small (less than three feet) plants coming back, a prescribed burn can be the cheapest way to get rid of those small cedars (see chapter 15).

Mechanical removal. Most folks choose mechanical methods for cedar removal, and there are several different options, all with pros and cons. Before you decide on a method, you should consider the following points:

- The first rule of brush control is to have a reason to remove the brush.
- The second rule is to know what will replace it—also when and how.
- The third rule is to know how the plant grows and what it takes to kill it.
- Know the characteristics and pros and cons of all brush control methods considered.
- Size matters. Methods appropriate for a five-acre ranchette may not be practical for 500 acres.

Brush control experts say that before you remove any kind of brush you should have one or a combination of the following reasons: fire protection, rangeland improvement, watershed improvement, or wildlife habitat improvement. The belief that cedar is simply bad and should be eliminated is not a good reason.

Many people make a fundamental mistake by assuming that, if they just remove something that in their view is bad, it will be replaced by whatever they think would be good. Well, it ain't necessarily so. Sometimes it will be replaced by something worse. For things other than cedar, sometimes root sprouts proliferate in abundance. For cedar, the bare areas frequently are initially covered with thistles and buffalo bur, and the thistle seeds may spread to other areas. In time, these forbs are usually replaced by grasses, but the initial visual impact can be quite depressing to those not expecting it.

As discussed earlier, Ashe juniper is the easiest brush pest to remove because it does not root-sprout. That does not mean that you won't have any more cedar bushes; lots more will come up from seeds, usually in spots where cedar bushes used to be, or under deciduous trees. These small cedars will have to be dealt with later, as discussed below.

If wildlife habitat is of value to you, some cedar should be left even on flat areas. Unless your only consideration is to maximize grass production, a cedar cover of 10–20 percent, especially in bunches thick enough to hide an animal, may be ideal. You may be able to select the space for this habitat that also provides for privacy or blocks an unattractive view.

For many people, the primary considerations in choosing a cedar removal method are cost and time. I would suggest two other factors that should be considered carefully: land disturbance and the likelihood of erosion, and damage to hardwoods. From a land management point of view, the latter two factors are more important and long-lasting than the former two.

For larger properties, bulldozers are probably the cheapest method, but they can also be the most destructive to the land. (Chaining, which involves two dozers with a chain between them, is suitable only for very large areas without mature hardwoods and is not discussed here.) Dozer treads tear up the surface of the soil, leaving bare ground and loosened soil to erode. The act of pushing the bushes into a pile for burning creates even more soil disruption. That aside, simply pushing over the bushes does not always kill them, but digging them up exposes a lot more soil to erosion and leaves the surface very uneven. Additionally, these large machines are not easily maneuvered and often run into hardwood trees, knocking off bark and exposing oaks to oak wilt. If dozers must be used, the smaller the better. Dozers have some advantages over other methods for other brush species, but for cedar I suggest other methods. The same thing goes for the large track-hoe machines that also dig up the bushes.

Probably the most common machine used to clear cedar in the Hill Country is a skid-steer tractor (Bobcat) with hydraulic cedar shears or snippers. These machines are much smaller and lighter than a dozer and have rubber tires (although in mesquite and prickly pear country some operators put metal tracks over the tires to prevent punctures) that are less disruptive to the soil. They are driven into a cedar tree and cut off the tree at the ground. They can then pick up the tree with the shears and carry it to a pile for burning. Since they are smaller, lighter, and more maneuverable, there should be less damage to the surrounding hardwoods, although some operators are not as careful as they should be.

Another machine-based method is a shredder or mulcher, which is a device on a large skid-steer tractor or an even larger articulated machine that can grind up a cedar bush where it grows and turn it all into a very coarse mulch (1–2 feet long and 1–3 inches in diameter). These machines have both advantages and disadvantages over, say, the hydraulic shear method. Their main advantage is that the mulch is spread over a wide area, so there is no brush to have to pile up and burn. The mulch covers the ground and protects it very well from future erosion and can also serve as a protective cage for new grass or forb sprouts to get established before they are grazed. The main disadvantage of these machines is that, if the cedar trees are large and thick, the process can generate a lot of mulch—enough to cover the ground 6 inches to a foot thick. This makes it difficult for grass or forbs to come up through and for animals or humans to walk over, and, being cedar, it doesn't decay very fast. It might be possible to rake up the mulch from the thicker places with some machine, but I haven't seen it done.

Finally, the next-most common method, and the one that is clearly best for the land, is the chain saw. It is much slower and somewhat more expensive (if you are hiring it done) than the other methods, but it is far less disruptive of the soil and also allows for several other beneficial practices. It is much easier to be selective when cutting cedar by hand—taking this bush, leaving the next one—being careful to leave a rare madrone or native cherry tree coming up inside the cedar. Cutting by hand also allows for assessment after each tree to see if you are beginning to cut down your visual barrier between you and your neighbor. In addition, it allows you to assess the shape of a tree and perhaps decide to cut the lower branches to make it into a "tree shape" rather than remove it entirely. Basically, cutting cedar by hand allows you to be sure you are doing what you want to do and makes it less likely that later you feel you have taken too many trees. With hand clearing, it is much easier to go back later and remove some trees that were left the first time, especially if you are doing the work yourself. Perhaps the best thing about hand clearing is that it allows you to do things with individual limbs that are beneficial for the landscape and habitat (see below).

What to Do with the Cedar (To Burn or Not to Burn)

Once you kill the cedar by any one of the above methods, you have in theory the following options: (1) leave it where it lies, (2) pile it up and leave it, (3) pile it up and burn it, (4) chip it into mulch, (5) put the branches to various beneficial uses, or (6) haul it off. Most people wind up with (3), so let's start there. What is wrong with burning?

There are several problems, or potential problems, with piling cedar up and burning it. First, inexperienced landowners and hurried contractors conduct a lot of brush pile burns under inappropriate conditions and either damage adjacent hardwoods or have the fire get away from them. Also, burning large piles heats the ground beneath the fire so much that it destroys the organic matter, including seeds, leaving a large circle of sterile soil that can take years to recover if left alone. Finally, although it is not often discussed, there is the issue of air pollution.

There is a wrong way, a good way, and a better way to burn a pile of cedar. The wrong way generally involves making a very large pile, not necessarily that far from uncut cedars or hardwoods; paying no attention to the weather (relative humidity and wind speed), current or forecast, or to the condition of the pasture or

trees surrounding the pile; lighting the pile at three in the afternoon; and with the most part burned by five, leaving to drive back
to Houston. It happens. Another wrong way is to tell the contractor
who is doing the clearing to pile it up and burn it, and then when
you next visit the place and find it not burned, to call up the contractor and berate him for not getting it done and insist he burn it
next week, even if it is the hottest, driest August on record. That
happens too. No matter what or when you want to burn, you must
observe county burn restrictions and red flag warnings.

A good way to burn cedar is to make moderate-size piles well
away from other cedar or hardwood trees, wait for a cool, damp
day, possibly even during a drizzle or light rain, when the wind is
light and predicted to stay that way and there is no forecast of a
cold front. Then clear away all fuel from around the pile, including grass, or wet it down. Then light the pile in the early morning,
with water, a shovel, wet burlap sacks, or similar equipment and a
cell phone on hand. Watch the fire until virtually all of the cedar is
gone, and then douse the coals with plenty of water before leaving
in the afternoon.

The best way is to pay attention to the surroundings and weather, as above, but also to make the cedar pile several feet away and
upwind from where you are actually going to burn. Put a few limbs
in the place you want to burn, start the fire and then feed it slowly
with one limb at a time so that you have complete control of the
size of the fire, even if the wind does come up. Refrain from putting the largest logs on the pile, because they burn the longest and
are the hardest to put out (set them aside and make small wildlife
habitat piles with them). By keeping the fire small and feeding it
slowly, you have not only complete control but also a cooler fire
that will not damage the underlying soil as badly.

Cutting the cedar and leaving it where it falls may not be a very
aesthetic way to leave the land, but it is better than one might
think, for it creates protected areas around the original cedar tree
where new grasses, forbs, or other trees can grow and where the

limbs can protect the soil from erosion until the new vegetation takes hold. If you plan to do a prescribed burn in the future, this method is effective and the burn will eliminate much of the cedar limbs then. Piling it up and leaving it may make for a good small animal habitat, although it would represent a huge fuel load in a wildfire or prescribed burn.

If you have a relatively small property, chipping the cedar limbs into mulch has some real advantages in that it reduces the volume manyfold and the chips can be used on flower beds, gardens, and paths. If the cedar has been cut by hand so that each limb was cut off the main trunk, the individual limbs can be fed into a relatively small chipper if the big branches are trimmed (we have an 18 HP chipper that can be pulled by a garden tractor and can handle limbs up to about 3.5 inches). This is certainly somewhat more time consuming than burning, but not as much as you might think if you would otherwise use the best method above for burning.

One of the best reasons to cut cedar by hand is that you get all of these cedar limbs with which you can do so many things to enhance the habitat and protect the soil. I always lay a few limbs down around the area where the cedar brush was to hold the leaves and litter in place so no bare soil is exposed. These limbs also act as nurse areas where new vegetation can get a start before being grazed or browsed. I also lay limbs or trunks across the slope to make small dams to slow overland flow during rainstorms. The leaves and litter that collect behind these cedar posts eventually accumulate some soil, and in time this becomes good soil for new vegetation. Limbs can be placed over any bare spots in the pasture to protect the soil and encourage new grass growth. And one can make a "teepee" from a collection of limbs stacked with the big end up and the small branches down to make cover and nesting areas for some birds and small animals.

If you are really determined to cut cedar from steep slopes, doing it by hand and laying the branches across the slope throughout the whole area is the best thing you can do. It is also advisable to do this in small patchwork areas at a time over several years so that the newly cleared areas are small. This does not completely solve the problems of working on a slope, but it helps mitigate them.

Finally, of course, if you have a place to haul the cedar—to a community burn center or chipping facility—that can be very helpful also.

Obviously these methods are not mutually exclusive. If you have individual limbs to work with, you can use some of them for the things suggested above and then burn or chip the remaining, smaller number at a later time.

Now That I Have Cleared the Cedar, I Can Relax, Right? Wrong

Unfortunately, the cedar will come back. But it is now a different and easier problem. Beginning about three years after the original clearing, you may begin to see very small cedars cropping up around the pasture, and as the years go by their number and size increase. It may take seven to ten years before these regrowth cedars reach three feet high and an inch in diameter, and it is before they reach this size that they are easiest to remove. If goats are ever to be effective for cedar control, it is at this stage. These small bushes are also the easiest to kill with herbicides.

Most people resort to either mechanical methods or fire to control these small cedars. Mechanical control means cutting the bushes down with loppers, pruners, or some sort of hand-held cutter blades. As long as the cut is made below the lowest green branch, the bush will die. Since it takes several years before these regrowth cedars get very big, there is lots of time to do the job.

Fire is what controlled the cedar before the area was settled, and prescribed burns are the cheapest, most effective method to maintain control of regrowth cedar. This should be done before the cedars are greater than three feet tall to be really effective (see chapter 15).

As the years go by, you will see fewer and fewer regrowth cedars, but it is never safe to assume that "the problem is solved," because cedar management is truly an ongoing job for any Hill Country landowner. The good news is that cedar berries are viable for only about eighteen months, so they don't just lie there for decades waiting for conditions to sprout like some grasses, forbs, and mesquite.

Although removing cedar presents many challenges and cedar does have some redeeming features, don't be discouraged about reducing the amount of cedar on your land. We really do need to reduce the amount of cedar in the Hill Country for the sake of our dwindling habitat and water supplies. The point is not to avoid doing so, but to do so with a plan based on knowledge of all the relevant factors. Most landowners who have reduced the amount of cedar on their place in a systematic, planned way are very much satisfied with the result and think their land is the better for it. You may not get a new spring when you do it, but you will get a more diverse, healthier landscape.

10
Erosion

IF YOU ask most landowners what is the most valuable thing on their land, some say the trees, some the stream, some say their houses, and some may say their pet horses. But almost no one lists what is truly the most valuable thing they have: their soil. When people look over a prospective piece of property to buy, they may be looking for an attractive landscape, good trees, maybe a certain type of terrain, maybe a view, maybe a source of water, but have you ever heard of anyone looking at how much soil is on the property?

Without soil, you can't grow plants. Without plants, you can't raise animals—either livestock or wildlife. Regardless of how over-grazed or overbrowsed or cedar-covered a landscape is, if there is soil, it can be restored to good health with time, effort, and a little money. Without the soil, restoration is virtually impossible, even with a lot of money; to haul in six inches of soil, it would take roughly fifty dumptruck loads per acre.

The natural processes of rocks breaking down to make more soil occur at a rate of about the thickness of a dime every hundred years. The point to all of this is to emphasize that your soil is so valuable that every effort should be made to hang onto what you

Eroded streambank exacerbated by livestock trails, resulting in loss of vegetation and soil.

have. Your neighbor may be happy to get the soil that washes off your property, but you should not be that generous.

There are two types of erosion: sheet erosion and erosion of creek, stream, and dry wash areas. Some erosion along draws and creeks is normal and has probably eroded most of the soil away already, leaving a largely rocky streambed. In places where there is significant erosion of steep banks along creeks, mitigation of those kinds of areas is beyond the scope of this book and the expertise of the author. In these areas where creeks are undercutting soil banks, the erosion is quite visible and seems to cry out for some action, but in many cases the cure is worse than the disease. Management of riparian areas in general is the topic of chapter 14. The discussion here concentrates on erosion from the land in general, both relatively flat areas as well as slopes. When water flows across the land carrying soil with it, it is called sheet erosion.

Sheet erosion can happen whenever rainwater is falling on the ground faster than it can soak in, and thus some of it has to run off. If it carries any soil with it, then erosion is occurring. Therefore, anything that can be done to stop the water from running off, or, failing that, to stop it from carrying any soil away, is in the direction of goodness for preventing erosion. Simple physics dictates that the slower the water is moving, the less sediment it can carry. So anything that can *slow down the flow of water* can help prevent soil from being carried away. Likewise, anything that *reduces the volume of water* flowing across the land also reduces the amount of soil that can be carried away. And finally, anything that can *prevent the soil from being dislodged* and becoming suspended in the water also helps reduce erosion.

So the strategy for preventing erosion is to get as much rainwater as possible to soak into the ground, thus reducing the volume of water flow, then to slow down the flow as much as possible so the water can carry only very small amounts of sediment, and at the same time doing whatever is possible to prevent the soil from being dislodged and suspended in the first place.

Fortunately, there is one action that accomplishes all these strategic elements at once: *grow more grass.* As discussed in chapter 7, the growth characteristics of native grasses are such that the root system with its accompanying microorganisms makes for a very porous soil under the grass plant, which absorbs water at a much faster rate than does bare ground. In this respect, grass is generally better than forbs or woody plants. Thus, rain falling on grass is more likely to soak into the ground, thus leaving less water to run off, than rain falling over any other surface. I don't want to leave the impression that forbs and woody plants contribute nothing to prevention of erosion, because they can be important. Any vegetation is better than nothing. And some woody plants grow in a fractured area where the cracks in the limestone can become a conduit for rainwater into an aquifer.

As the water does begin to flow over the ground, it encounters first one grass plant base, then another, along with the usual dead leaves lying among the plants, and thus it is slowed down, dropping some of the sediment. This also gives the water more chance to soak into the ground.

A raindrop falling from the sky hits the ground at about 20 miles/hr, which is enough force to dislodge particles of soil (as evidenced by the brown stain on the white stone of some newly built houses). The particles can then be easily carried away by the water. But a good stand of grass has a dense enough cover of leaves, green and dead, that protects the soil beneath from the falling raindrops and thus prevents the soil from being dislodged.

Aside from managing your land for a good stand of native grasses, other things you can do involve causing as little disturbance as possible to the surface and mitigating any disturbance you do make. Recovery of disturbed land is much slower in the Hill Country than in areas of deeper soils and more rainfall. This generally means limiting any vehicle traffic to the absolute minimum necessary, especially ATVs and dirt bikes, which are really destructive; choosing the least disruptive method you can for any brush clearing; keeping livestock from trampling bare spots around feeding and watering locations; and in general being cognizant of the fact that the land is fragile and recovery is slow.

One of the benefits of cutting cedar by hand (see chapter 9) is that it generates many cedar branches that can be used to great advantage to mitigate potential erosion spots. Lay a single layer of the branches over bare places (including those where the cedar once grew) and across the slope to help protect bare soil and to slow the rate of runoff. When cedar logs are laid across the slope, they soon catch cedar needles and other leaves, followed by sediment, and this mixture becomes a rich soil for new grass and forb plants to sprout from. The cedar branches also protect newly sprouted grasses, forbs, and woody plants until they are big enough to rise above the branches and are thus better established to withstand grazing/browsing.

Rocks are also effective in slowing water flow, especially in areas where the flow accumulates into small streams. Of course you can't fight the terrain. If the contours of the land funnel water to a particular place, you are not going to change that, short of a lot of dirt work, but you can try to slow down the flow so as not to lose so much soil. In general, anyone who observes water flow across their land can come up with ways to slow it down.

If you don't think you have erosion on your place, I suggest you take a walk around during or just immediately after the next two- to three-inch rainfall and watch the water move across what is supposedly flat land. Look too for small bunches of grass that appear to be growing up on a little pedestal (it is called "pedestaling") or rocks that seem to be sitting up on a small mound of soil. Neither the grass nor the rocks looked for a high place to be; rather, the soil around them eroded away. Anywhere you find this condition you have already, and probably still are, losing soil. What you need to do is improve the grass cover and work to slow down the flow of water across the land.

One of the greatest scars on the beautiful Hill Country landscape is the result of road and house-pad construction, where large amounts of earth are scraped off the upslope and piled on the downslope to make a level road or homesite. The loosened, yellow caliche mixture then becomes riddled with erosion gullies, and the lighter material (largely soil) is deposited somewhere downslope, frequently on the neighbor's property. There are ways these scars can be prevented, minimized, or mitigated, but all too often no effort is made to do so. What is needed more than anything else are knowledgeable, concerned landowners with a desire to do the right thing.

I I

Deer

WHITE-TAILED deer are primarily browsers, eating mostly woody plants and forbs. Under normal conditions,
they consume less than 15 percent of their diet in grass.
Since they are ruminants like cattle, sheep, and goats, one might
ask why they don't eat more grass. The answer is that, compared
to cattle, deer have a relatively small, open rumen (fore-stomach)
with a low residence time (six to seven hours vs. about twenty-four
hours for cattle). The consequence of the shorter residence time of
food in the rumen is that the more difficult-to-digest foods, such
as those with a lot of lignin like grass, are not very well digested.
As Bill Armstrong, wildlife biologist of the Kerr Wildlife Management Area, often says, "Deer will starve to death with a belly full
of grass." Thus, tree leaves and forbs, which are easier to digest and
more nutritious, are their favored food. Deer need four pounds of
dry food, containing 12 percent or more protein, per day.

We tend to think of sheep and goats and the exotics as being in
competition with deer for food, and in fact they are to a large extent, but all of these other animals have an advantage over white-
tailed deer: they can all eat grass and survive (see the experiment
with white-tailed and sika deer in chapter 8).

The peak breeding time for white-tailed deer in the Hill Country is around Thanksgiving, November 17–20, but breeding goes on from late October until early January. The gestation period is 6.5 months, and 90 percent of the fawns are born by June 30. On average, does have 1.7 fetuses in midwinter. In ideal habitat twins are common, but in most of the Hill Country does usually raise at most one fawn to maturity. One fawn per doe is referred to as a 100 percent fawn crop. In highly overpopulated areas, such as most of the Hill Country, it is not uncommon for the fawn crop to be 30 percent or less. The reason for the low fawn numbers is mainly poor nutrition of the does, and possibly also poor hiding cover for the small fawns.

We commonly think of deer society as one in which the biggest, strongest buck dominates the herd, driving off the lesser males and mating with all the does. DNA evidence recently has shown that in fact, where does have two fetuses, they frequently have two fathers, and that not nearly as much of the breeding is done by the larger mature bucks as previously assumed. It seems that in deer society there are lovers and there are fighters, and they are not necessarily the same.

Experts say that the carrying capacity for white-tailed deer in the central part of the Hill Country is no more than a deer to every eight acres, and most would rather see that number be a deer to every ten to twelve acres. The Hill Country average is more like a deer to every three to four acres, and many places have even higher populations, where the animals are obviously malnourished.

Why has the population grown so high? There are several reasons. It began with the early settlers who overgrazed the land and reduced fires, thus converting grasslands (poor deer habitat) into savannas (good deer habitat) and, perhaps more important, reducing and eventually eliminating most big predators. Early hunting regulations limited the number of bucks that could be taken and outlawed taking does. The elimination of the screw-worm fly in the 1960s removed the last significant predator of the deer other than human hunters and drivers. Today, hunters take an average of only 1.1 animals, in spite of a license to take five, and even now not as many does are harvested as should be for best population control. So, effectively, the population of deer is limited by the food source, and that limitation takes place only after the supply has been exhausted and the animals are malnourished.

Controlling Their Numbers

I start with the recognition that people like to see deer and that no one seeks to eliminate them. Still, too much of a good thing can be bad, like the effects on the habitat associated with overbrowsing the landscape (see chapter 8). The ideal situation would be to have a white-tailed deer population equal to, but no greater than, the number that can be sustained on the land and still maintain a healthy, biodiverse, sustainable habitat. In other words, to have the stocking rate equal to or less than the carrying capacity. Under those conditions, the animals would be healthy and well fed.

There are several complicating factors involved in controlling numbers; here is a partial list:

- You probably don't know exactly how many deer you have on your property.
- The white-tailed deer may not be the only animals eating certain plants; goats, possibly sheep, and all of the exotics get the blame too.
- All white-tailed deer in Texas belong to the State of Texas, not to the landowner, so there are many rules and regulations about what you can and cannot do to/with them.
- The deer are mobile, moving between your and your neighbors' properties more or less at will, so even if you do your part to control numbers, if your neighbors do not, your efforts may be largely ineffective, especially on smaller properties.

- Many people (neighbors, relatives, family members, maybe yourself) have serious problems with the idea of killing deer, which makes control much more difficult.
- Even though you can have as many hunters hunt on your property as you wish, and each one can take as many as five deer, the average hunter takes only 1.1 deer (freezer space limited).

For all of the above reasons, many people fail to reduce the population of white-tailed deer to the desired numbers.

Why is it necessary to know how many deer you have? The answer is that if you have more than one deer to every eight to ten acres, in order for the habitat to improve you need to reduce the numbers. But if you don't know how many you have, then you can't know how many to take to bring the numbers into line with the carrying capacity. Many folks simply take a few deer, either themselves or their friends and relatives, and pat themselves on the back for having "done something." They may even do this for several years, assuming they are "controlling the deer"—without any evidence that they are in fact having any impact on the population. Most folks are surprised to learn that to reduce the deer population measurably it is necessary to take at least *30 percent* of the population, every year. The reasons for this have to do with the population dynamics of the herd, number of fawns raised to maturity, age distribution and lifespan, and other factors.

At any rate, the point is that if you have a deer to every three acres and wish to reduce the number to a deer to every ten acres, on 100 acres that means reducing thirty-three deer down to ten. That means taking about ten deer the first year, some of which will be replaced, then 30 percent of next year's population, and so on until you begin to get near the goal of ten deer. Then, to maintain the population at about ten deer, you will need to take 20–25 percent of the population every year, or two or three deer per year from now on. All of these numbers are estimates and depend somewhat on the age and sex ratio of the herd and of the animals taken, and they

This picture of a healthy spring fawn belies the difficulty it may have later in life finding sufficient food and the damage it may cause to the ecosystem's mid-level vegetation.

also assume that immigration and emigration are equal, which may not be the case, especially if your habitat begins to look better to deer than your neighbors'.

Counting Deer

Before you can make any plans to manage your deer numbers, you first need some estimate of the numbers you have to begin with, which means taking a census. The basic procedure for estimating deer populations is called a spotlight count. It works like this. Select a path or road that runs through your property encompassing all the different types of areas, so that it is representative of your whole ranch (not including feeders or watering areas). Measure the length of the path and then walk or drive it, estimating at every 0.1 mile the distance on each side of the path that you would be able to see a deer if it were there. Average the total sight line width and multiply that by the length of the path to calculate the number of acres you will be able to see deer along that path. Then, after dark, drive the path very slowly, search both sides of the path with strong spotlights, and count the deer you see. Divide the total number of acres searched by the total number of deer, and you get the density of deer, or number of acres per deer. Divide the total number of acres on your property where deer have access by the number of acres/deer, and you get the total number of deer on your property. Do this at least three different times and average the results. This will get about as accurate a number as is possible to obtain, but it is only an estimate. These counts are best done in August or September.

This method obviously works best on larger properties, where ingress and egress account for smaller portions of the population. It also probably works best in a vehicle rather than on foot, since the deer are more accustomed to vehicles than to people on foot and are thus less likely to hide. This method is not appropriate for twenty acres or less. On smaller acreages, one usually tries to count all of the deer on the property by going to various places with a spotlight. The problem is that the very act of counting may scare the deer, and they may jump the fence and not be counted. You could also walk a path that allows you to see most of the property and count the deer in the daylight. Any of these methods should be repeated several times at different times of day or night. Also, don't just count animals around a feeder, or you're likely to get an unrepresentative sample.

If you have so many deer that they are up and foraging all during the day and they look skinny, you probably have a density of more than one deer to every two acres. Unless your property is in pristine condition, with tree leaves on branches hanging down to the ground and many edible forbs, you probably have more than a deer to every eight to ten acres. In more intermediate conditions, then, you should come up with a number somewhere between a deer to every two acres and a deer to every eight acres.

Another way to estimate deer populations is to observe what is and what is not eaten. This won't give you a number, but it will tell you what the animals on your place are doing to your habitat. Wildlife biologists observe that deer should not be eating live oak sprouts or leaves from the live oak trees until at least December. If they are resorting to eating this less-than-favorite food, then your land is being overbrowsed. If a walk around your pasture doesn't turn up any edible forbs in spring or early summer, or any woody saplings or root sprouts, or any tree branches (other than cedar) with leaves below the browse line, then you are probably overbrowsed. It is hard for beginners to assess their property this way, but, with practice and time spent on other properties and comparing those places to your own, you will begin to get the hang of it.

Everything I have said about counting deer also applies to any of the exotic ungulates or your neighbor's goats if they habitually get into your pasture. And if they are a problem, until you get some control of their numbers, your efforts to control the deer will be ineffective in improving the habitat.

Removing Deer

There are two ways to remove deer from your property: you can shoot them or, with a permit, trap them. Shooting is obviously the cheapest, easiest, and most selective method to reduce the number of deer. If the number to be taken is within the capability of the number of hunters you wish to have on your property (i.e., five deer per hunter), then all that is required is a hunting license. If the number to be taken is in excess of what can reasonably be taken by hunters, it is possible, working with the Texas Parks and Wildlife Department, to obtain a permit to take more deer than a regular hunting license allows. To do so requires a wildlife management plan and a census, usually developed with assistance from a TPWD biologist, who will also provide a target number of animals to be taken. These permits are issued to those landowners who need to reduce the number of deer significantly because of the effect they are having on the habitat, and the landowner has some responsibility to keep harvest records and to manage the land in a way that will in fact improve the habitat.

It is also possible to get a permit from the TPWD to trap the animals under certain conditions. There is one such permit that allows for trapping the deer, transporting them to another location, and then releasing them. It is extremely difficult to obtain such a permit because the property on which they are to be released must be approved by TPWD and have a very low population of existing deer, and there are very few such properties anywhere in the state. The other trapping permit allows trapping of deer and transporting them to a facility where they can be slaughtered and processed and the meat donated to charity.

Trapping deer, usually done by a remotely operated drop net trap, is expensive and time consuming, and if the people doing the trapping are not experienced many animals can be injured or killed in the process. Some people feel that trapping is more humane, but in fact the animals are under extreme stress during the whole process, and if they are to be killed anyway, it is hard to imagine that we are doing them any favors with such traps.

Whatever the method you use to control your deer numbers, there are three points you must bear in mind: (1) To reduce the number of animals, you need to take at least 30 percent of the population per year; to maintain the population at its current level, you need to take between 20 and 25 percent a year. (2) Controlling deer populations, like controlling cedar, must be an ongoing operation, not something you do once and consider it done. (3) Removing does from the population has a greater effect on the population than removing bucks.

A deer standing "knee deep" in grass, obviously malnourished for lack of browse or forbs.

High-Fencing

Building high fences (seven feet or more) became popular when people began to raise expensive, exotic animals, and even though high fences are expensive they are the only economical solution for ranchers wishing to raise exotics. These fences are designed to keep animals in, and some ranchers are trying to keep their better white-tailed bucks from escaping as well. Other folks who contemplate high fencing, but who do not own exotic animals, are trying to keep animals out. There are fundamental problems with either purpose.

Since these large ungulates have few natural predators in Texas, animals that are fenced into a given area have only two controls on their numbers—food availability and human activity. If we do not control their numbers adequately, the animals eat all the available forage and damage the habitat, sometimes irreparably. High-fencing a property means that one must control the numbers within from now on.

High-fencing to keep animals out has a different set of problems. Doing so interrupts the normal movement of many different animal species, possibly preventing some from getting to their usual water or food source. If too many landowners put up high fences, these barriers become real impediments to normal movement and may even wind up trapping animals inside a collection of high-fenced properties. Furthermore, if there are no animals inside such a fence, then the plants are not browsed or grazed as they naturally would be and in time can form impenetrable thickets. In short, high fences can alter habitats, and not always in ways that are anticipated or wanted.

Of course, fencing to keep animals in or out are not two mutually exclusive conditions, because every high fence really does both. The main advice that can be given to anyone contemplating high fencing is to take into consideration both the pros and cons as well as alternative practices before doing anything. Also, if you use high fencing, monitor what is going on inside your fence closely year after year; you have greater than normal control of the habitat inside the fence, so it is up to you to manage it wisely.

To Feed or Not to Feed

Lots of people like to feed deer, some because they want to tame them and be able to see them up close and often, and some because they feel sorry for them. Some even look forward to this well before they move to the Hill Country. So the subject is a highly emotional one, and I have no intention of taking sides on this issue.

I can tell you that the vast majority of wildlife biologists would advise against feeding wildlife, especially deer. But then these same folks usually think feeding birds is okay. Biologically, feeding deer on a regular basis certainly increases the population in the vicinity of your feeding area, which brings more browsing to the area than would otherwise occur. Clearly, then, the practice causes the habitat to be degraded somewhat more than if the deer were not attracted to the area. How degraded depends a lot on how many animals are attracted to the area on a regular basis.

If you are feeding them deer pellets and this is their sole source of food, you need to give them about four pounds of pellets per day. At that rate, if you are feeding six deer, you will go through nearly a 50-lb bag every two days. Most people don't feed anywhere near that much, so the deer still have to browse for food.

If you are going to feed deer, you should at least give them something that is good for them, meaning deer pellets and not corn. Corn is primarily starch. It is low in protein (about 7 percent) and can have deleterious amounts of aflatoxin, a potent carcinogen. Pellets are high in protein (20 percent) and can actually help the deer in terms of their ability to carry fawns and grow antlers. Note that I am referring to long-term feeding throughout the year. The baiting with corn of certain areas during hunting season is just that, baiting, and does not contribute much to their overall nutrition.

Obviously, the fewer deer you feed, and to some extent the less you feed them, the less you interfere with the natural population and browsing. But humankind has already greatly altered the natural population of white-tailed deer by eliminating predators and providing habitat with our landscaping. I happen to believe that in the best of all possible worlds the deer populations would be significantly lower and within the carrying capacity of the habitat. Under those conditions, the sighting of a deer would be an unusual event and a joy to see rather than a common sight. Also, I don't have to be able to see a wolf in Yellowstone to feel good about the fact that they are there and healthy and living the way nature intended. Similarly, if I saw deer less often I would feel good knowing that they were hiding and resting during the day with a full stomach.

Allow me one final point about deer, a pet peeve of mine: Deer don't have horns, they have antlers. Horns grow continuously like fingernails throughout the life of the animal. Antlers are shed every year and regrown.

12

Oak Wilt

ANY sources of information on oak wilt are available. The two best are a pamphlet from the Texas Forest Service, "How to Identify and Manage Oak Wilt in Texas," and the website www.texasoakwilt.org. Because these are complete, factual, and updated periodically, I just provide a brief introduction here and don't belabor the topic.

What It Is

Oak wilt is a disease that affects oaks. It is caused by a fungus that gets into the vascular system of the tree and plugs up the channels that transmit water and nutrients from the roots to the leaves and carbohydrates from the leaves to the roots. It is similar to Dutch elm disease. All species of oaks can be infected with this fungus. White oaks (shin oak, Lacey oak, post oak, chinquapin oak) are seldom seriously affected, so if they were to be infected you probably would never know it. Red oaks (Spanish oak/Texas red oak, blackjack oak) are highly susceptible and frequently die within a few weeks. Live oaks are intermediate in susceptibility; about 80 percent will die, although it may take months, and the others may be damaged (greatly reduced canopy cover) but survive.

The fungus can be transmitted in two ways. If an infected red oak dies of oak wilt, especially late in the year, the next spring the fungus may form a fungal mat in spots underneath the bark. This mat contains spores of the fungus and it may swell enough to crack the bark, which allows tiny sap beetles, attracted by the smell of rotting fruit, to the site where they feed on the fungus. The beetles get the spores on their bodies and then fly off in search of a new source of sap. Often, the most likely source of sap is a fresh cut or wound on another oak. If the beetle finds such a site, spores are transmitted to the second tree and the fungus again grows in the vascular system of that tree, thus starting a new oak wilt center.

The second way the fungus can be transmitted to other trees, and the way most live oaks get the disease, is by traveling through the root system. Most live oaks growing near one another, especially if they are of similar size, are probably root sprouts from the same mother tree (maybe now long gone) and are in fact really clones of each other connected together in the root system. Even live oaks that started out as totally separate trees can have their roots come together and graft so that the fungus can travel through the roots from tree to tree. Red oaks are much less likely to be connected via the root system.

This is why, even though live oaks are actually less susceptible to the disease than red oaks, we lose many more live oaks than red oaks; the former tend to be connected via the roots and, when one tree gets the disease, several do. Of course, we also have a lot more live oaks than red oaks in most areas of the Hill Country. When a red oak gets oak wilt, it may die very quickly, but it is likely to be the only tree in the area to do so. A new oak wilt center starting in a live oak may eventually spread some distance and infect dozens or even hundreds of trees.

What to Do about It

There is no cure for oak wilt. Repeat: There is no cure for oak wilt. There are preventative measures. To prevent a new oak wilt center

A telltale sign that oak wilt has visited the area, with a large number of live oaks killed as the fungus spread through the interconnected roots.

from getting started (by that sap beetle with spores on its body), you need to make sure that the bug doesn't find any open wounds or cuts where fresh sap is oozing out. The way to do that is to *always paint all wounds or cuts on all oaks in all seasons immediately.* Sure, limbs will be broken off in wind storms that you won't know about until it is too late for painting to do any good, but statistically most new oak wilt centers occur around human habitation, power lines, or roads where people are causing most of the wounds.

Once oak wilt is observed in nearby live oak trees, there are two ways to prevent it from spreading to other trees in the area. One is to trench between the infected tree(s) and apparently healthy trees, as long as this can be done at least 100 feet away from the infected tree(s). The trenching needs to be deep enough to cut all possible connecting roots, which depends on the depth of the soil. This severs any possible root connection, thus preventing the fungus from passing from one tree to the other. The other procedure is to inject any healthy trees that are 100–150 feet from an infected tree with the fungicide Propiconazole 14.3. This is the only agent proved to be effective in preventing the disease from infecting a tree.

The details of both trenching and fungicide injection are discussed on the above-mentioned website. It is important to emphasize one thing about the fungicide treatment. It is about 80 percent effective if the treatment is done at the optimal time. The optimal time is when the disease appears to be approaching the tree to be treated but has not yet arrived, which usually means that the disease is affecting trees no closer than about 75 feet and no farther than about 150 feet away.

Why all the concern about timing? Of course, we can't see the fungus progressing through the roots, but think of it as a slow-moving grass fire. It won't do any good to spray water on the grass too far in front of the fire, because it will just dry out again before the fire gets to it. But if you wait too long and the fire is too close, you may not be able to spray water fast enough to put it out. The fungicide treatment places the fungicide in the active vascular system in the cambium layer and kills the fungus when it tries to grow up the same vascular system. If you treat years before the fungus arrives, the tree will grow new cambium layers and the fungicide will be trapped within old channels, not where the fungus will be growing. If the fungus has already reached the tree, it may have already started to plug up the channels, so the fungicide may not be able to reach it all to kill it. The fungicide is simply a poison for fungus; it does not give the tree immunity. (Plants don't have immune systems; they cannot be given vaccinations and made immune to anything.)

How to Tell If You Have Oak Wilt

The pamphlet and the website cited above have good pictures of the symptomatic leaves usually, but not always, seen on live oaks with the disease. If you see live oaks dropping their leaves at any time other than the normal leaf-exchange time of late March to early May, and you find leaves with the "fish-bone" pattern seen in the photos, you probably have oak wilt. Also, if you have several live oak trees in close proximity that seem to be dying and losing their leaves, that is a very good sign that you have oak wilt. When in doubt, try to get someone from your local Texas Forest Service office to come out and look. Red oaks don't necessarily display any symptoms that are unique to oak wilt, other than dying very fast.

Both trenching and Propiconazole 14.3 treatments are expensive, so you want to be sure you really need them, that the timing is right, and that the work is done correctly. Getting several opinions, especially from someone not in a position to gain financially from giving the advice, is a good idea. At the same time, there are a lot of "homemade remedies" that are useless or even harmful. Know that the fungus is not in the soil and that, although Clorox may kill fungus in the bathroom, pouring it on the soil won't do anything to the fungus but may damage roots.

Will oak wilt eventually wipe out all the oaks in the Hill Country? No, most of the white oaks and even some of the live oaks will survive. The red oaks (Spanish and blackjack) would probably survive if young replacements were allowed to grow to mature trees, but given the deer and exotic population, plus losses to wind and other diseases such as hypoxylon, they are indeed in peril. As live oaks get less and less common, the isolated ones will have a much better chance of not contracting oak wilt. Whether oak wilt or the lack of young replacement trees will have the greatest effect on the live oak population is debatable right now. All of this argues for everyone planting lots of native trees, of all species, including live oaks and Spanish oaks.

13

Exotics

F OR the purposes of this discussion, an exotic can be defined as an introduced species not native to this area or to surrounding areas. It can be an animal (including insects) or a plant, and its introduction may have been intentional or accidental. It can be something introduced many years ago (e.g., bermudagrass in the 1700s or English sparrows) or something introduced recently (e.g., Asian carp, zebra mussels).

The majority of these foreign species cause no harm to us or the environment. All of our common livestock (horses, cattle, sheep, goats, pigs, and chickens) and most of our vegetables and food grains are in fact introduced from other continents. Our use of these plants and animals has of course had a huge impact on the current ecology of the Hill Country, but most of the impact comes from people altering the landscape in order to provide food and fiber for themselves. Had there been native species that served the same purposes, the same result would have occurred.

Sometimes, however, the introduction of exotics causes great disruptions in the native flora and fauna. This happens whenever the introduced species escapes our control and begins reproducing and spreading into places where they were not intended to be. Feral hogs, kudzu, and fire ants are three such examples. The introduction of a new species into a healthy, balanced ecosystem can be quite detrimental for several reasons: (1) it can crowd out one or more native species, and the native species may have been a food source for other natives; (2) its growth habit may form thickets unsuitable for wildlife; (3) it may use up natural resources needed by natives; and (4) whatever once controlled the spread of the exotic in its native land may be missing here, so the exotic expands uncontrollably.

Fire ants have competed with and reduced the population of native harvester ants, which are the main food source of the Texas horned lizard (aka horny toad), thus contributing to its decline. Feral hogs cause great destruction to fences, gardens, fields, and native habitats and have contributed to the decline of ground-nesting birds. Feral cats kill thousands of songbirds and small mammals every year. Salt cedar, vitex, and giant reed crowd out many streamside plants, producing dense, impenetrable thickets of poor monoculture habitat. Johnsongrass and KR bluestem (and others) compete too well with native grasses, taking over some areas once occupied by better-quality native grasses. Most of the introduced exotic ungulates compete with the native white-tailed deer for browse and forbs before switching to grass after all of the forage available to the deer is gone.

Most of the exotic land plants that are causing problems were intentionally introduced by the nursery trade to satisfy customer demand for something new, different, prettier. Many of the aquatic plants and animals were introduced as aquarium species, which were then dumped into our water systems. Almost all of the exotic ungulates were intentionally brought in as pets or for sport hunting. Many of the exotic mice and rats and most of the insects were introduced accidentally.

The problem with even the most careful introduction of a new species is that we are not smart enough to understand fully everything about the ecology and environment the species came from or how it will fit into the new ecology. So things that were completely controlled in their native environment may become uncontrolled

This blackbuck antelope is a symbol of the many exotic plant and animal species that now inhabit the Hill Country. Many of these species are much more problematic than the antelope, but all exotics have the potential to become invasive and do great damage to the local habitat if their numbers are not controlled.

and invasive here. Because we can't know the consequences of an exotic introduction into our ecosystem, and because the consequences can be so costly and severe, a far better policy would be to never introduce any nonnative plant or animal into the area, and to not buy any nonnative plant, of which there are many, for sale in local nurseries. Apparently well-behaved adaptive plants from adjoining ecosystems seem to present little risk.

In 1963 it was estimated that there were thirteen different species of large exotic animals in Texas, and over 13,000 individuals. By 2006 those numbers had escalated to over one hundred species and 250,000 animals. The most common of the intentionally introduced ungulates are axis deer (*Cervus axis*), sika deer (*Cervus nippon*), fallow deer (*Cervus dama*), aoudad (*Ammotragus lervia*), blackbuck (*Antelope cervacapra*), and the most destructive of all, feral hog (*Sus scrofa*). Other exotic mammals in our midst are nutria (*Myocastor coypus*), house mouse (*Mus musculus*), Norway rat (*Rattus norvegicus*), and feral cat (*Felis catus*).

Among the most common, serious insect pests are red imported fire ant (Solenopsis invicta), Asian tiger mosquito (*Aedes albopictus*), Africanized honeybee (*Apis mellifera scutellata*), Formosan termite (*Coptotermes formosanus*), and German cockroach (*Blattella germanica*).

Just a few of the long list of troublesome invasive plants are shown in the table.

Chinese tallow, *Sapium sebiferum*
chinaberry, *Melia azedarach*
vitex, *Vitex agnus-castus*
nandina, *Nandina domestica*
wax-leaf ligustrum, *Ligustrum japonicum*
bamboo, *Bambusa tulda*
Japanese privet, *Ligustrum japonicum*
water hyacinth, *Eichhornia crassipes*
Japanese honeysuckle, *Lonicera japonica*

pyracantha, *Pyracantha koidzumii*
common cocklebur, *Xanthium strumarium*
salt cedar, *Tamarix spp.*
musk thistle, *Carduus nutans*
giant reed, *Arundo donax*
johnsongrass, *Sorghum halepense*
bermudagrass, *Cynodon dactylon*
KR bluestem, *Bothriochola ischaemum var. songarcia*

The bottom line is that anyone interested in preserving the native habitat and native flora and fauna should avoid the introduction of any nonnative species of plant or animal and should work to eliminate any of the above listed invasive species seen on their property. That said, there are some species (e.g., fire ants and KR bluestem) whose eradication may be so expensive and likely to cause collateral damage (killing native species as well) that large-scale eradication efforts are not recommended. Simply repeatedly cutting down any of the above invasive woody plants (and perhaps treating the cut stump with herbicide) goes a long way toward our collective effort to eradicate these problem plants.

14

Riparian Areas

IPARIAN zones are special places that require special care. A riparian zone is the area along a stream or lake from the edge of the water up to and including the floodplain. The floodplain is the land adjacent to the stream that is subject to flooding. If you have such an area you are quite fortunate, for it represents a totally different ecosystem with different vegetation and wildlife from upland areas and acts as a wildlife magnet attracting many animals you are not likely to see otherwise. These areas can be quite robust and resilient, but they can also be quite vulnerable to severe degradation. It is for this reason that special care needs to be given to these areas.

The vegetation in a riparian zone is different from that of the surrounding countryside, largely because the plants can have their roots down into the local water table and thus have available much more moisture. These plants are also protected from most drought periods. Protecting the natural vegetation within the riparian zone should be the single most important management goal, because the health of the streamside vegetation affects not only the overall health of the immediate area but the health of areas downstream as well.

Streambank badly eroded from continuous use by cattle (see also facing page).

Vegetation frequently found in riparian areas includes sedges and rushes such as spike rush, emory sedge, and bull rush; grasses such as switchgrass, Eastern gamagrass, Lindheimer muhly, and bushy bluestem; forbs such as American water-willow and water primrose; shrubs such as buttonbush, indigobush (false indigo), elderberry, willow baccharis, and spicebush; and trees such as bald cypress, sycamore, pecan, and black willow. This is not to say that some upland species aren't found in riparian zones as well. What seems to be most important is a mixture of all of the above classes of vegetation, from the smallest rush to the largest tree, because this mixture of plant types is better at holding soil than any one

A healthy riparian area downsteam from the previous photo where the cattle have been fenced out.

type alone. And by far the most important goal of riparian zone management is to attain and maintain good healthy streambank vegetation to hold the soil in place even during flood events.

A healthy, vegetated riparian zone slows down the water during a flood and reduces erosion; slows down and catches sediment, thus improving the water quality; and stores water in the soil, which is then released slowly over time to maintain stream flow. Good streambank vegetation makes the stream itself healthier by not only filtering sediment but also shading the water and lowering its temperature. Healthy riparian areas are essential for flood control, water purity, and consistent flow. What happens on your land really does matter to your neighbors.

A healthy riparian zone is also an important wildlife habitat that is essential to birds, mammals, amphibians, and reptiles. Many of these animals can find food in other ecological zones, some far removed from the water, but they all need water, and thus this is where the greatest concentration of species can be found.

The problems with riparian areas are mostly the consequences of human activities: overgrazing, especially with no rest periods for recovery; mowing the vegetation along the stream; vehicle traffic along or in the stream, including ATVs and cars being washed in streams; construction sites too close to the riparian zone; attempts to redirect streams or alter bank topography and dam construction. Other problems include the invasion of various exotic plant species.

The most common problem with riparian areas in most of the Hill Country is overgrazing. It is common practice to use any available creek, free flowing or dammed up, as the primary water source for livestock. And since these areas are frequently flat and accessible, and adjacent areas are used for supplemental feeding, salt and protein blocks, and the like, the animals tend to spend a lot of their time in the area, not only grazing it heavily but also trampling the grass and creating trails. All of this activity gives rise to bare streambanks, which then become eroded. In addition, concentrating the number of animal-hours spent near the stream gives rise to pollution of the stream with animal waste, the effects of which can flow far downstream.

The problem is that for many properties the stream may be the only source of water for the livestock, and if that is the case it may be very difficult or inconvenient to provide other watering places and keep the animals away from the riparian areas. But in the ideal case, the riparian area would be fenced to keep the animals out except for specific times during the year when limited grazing could be beneficial to some of the vegetation, with water provided for the animals elsewhere, such as a water trough.

Where exotics appear in the riparian zones they should be removed if at all possible, since they can alter the vegetative makeup and diversity of the area.

The important moral here is that a healthy riparian zone is exceedingly important to the overall health of the whole area, including downstream areas. The health of the riparian area depends on maintaining a good diversity of native plants holding the soil, reducing storm flow energy, and filtering sediment. And healthy riparian vegetation can be achieved only by good management practices that protect the natural vegetation from overgrazing, overuse, and invasion by exotic species.

If you have such an area and are interested in further information on the subject, I recommend two publications: *A Texas Field Guide to Evaluating Rangeland Stream & Riparian Health* (2004), by Nikkoal Dictson and Larry White, Texas AgriLife Extension Service, publication B-6157, and *Streamside Management in the Hill Country: An Edwards Plateau Landowner's Guide* (2006), published jointly by The Nature Conservancy (www.nature.org/texas), NRCS (www.tx.nrcs.gov), Nueces River Authority (www.nueces-ra.org), and Guadalupe-Blanco River Authority (www.gbra.org).

15

Prescribed Burning

Game (wildlife) can be restored by the creative use of the same tools which have heretofore destroyed it—axe, cow, plow, fire and gun. The favorable alignment of these forces sometimes came about in pioneer days by accident. . . . Management is their purposeful and continuing alignment.

—*Aldo Leopold, 1949*

FIRST, let me make clear what we are talking about here. A prescribed burn is an intentional, planned burning of a pasture carried out under only certain weather conditions by experienced personnel after considerable planning and preparation. A prescribed burn is not any other intentionally set fire including those set to burn brush piles. Confusion exists because newspapers frequently report a wildfire as being caused by a "controlled" burn (even though if it got away it clearly was not controlled), and the public all too often confuses the term controlled with prescribed, giving rise to the common belief that prescribed burns frequently become uncontrolled. That does happen, but it is rare; properly done, prescribed burns are quite safe.

Why would anyone intentionally set fire to their pasture? There are several reasons: (1) It is the cheapest, easiest way to kill small (less than three feet) cedar bushes before they grow into larger, more problematic bushes. (2) If, in ungrazed areas, heavy thatch builds up to an extent that it prevents grass or forb seeds from germinating and begins to crowd out smaller grass species, then burning removes that excess thatch and rejuvenates the range. (3) The resulting ash contains a lot of nutrients, so the burn gives rise to a fertilization effect. (4) Burning frequently results in much greater vegetative diversity by stimulating many species that are not common before the burn. The bottom line is that prescribed burns help control brush and give rise to healthier, more diverse habitats.

Before proceeding with this topic, let me state that no one *should attempt to conduct a burn unless they have experience doing so,* which means having participated in the planning, preparation, and execution of prescribed burns conducted by experienced people. I would further suggest that no one attempt to burn a piece of property for the first time without getting a second opinion from an experienced person on how best to go about burning that particular area and having experienced help.

A question frequently asked is how is it possible to do this safely? The answer is that knowing the characteristics of a prairie fire, properly preparing the area to be burned, and strictly adhering to the proper weather conditions make it possible to contain a fire to the desired area. What many people don't realize is the close relationship between the intensity of a fire and the relative humidity; the lower the humidity, the more intense the fire. Most people do realize, however, that the higher the wind speed, the more intense the fire. Prescribed burns are usually conducted only when the humidity is above 20 percent and the wind speed is less than 20 miles/hr, usually less than 10–15 miles/hr. In addition, preparing the perimeter of the area to be burned and knowing how to set the fire relative to the wind direction make these operations safe.

One thing the landowner needs to know is that you don't just suddenly decide in late December to do a burn in January. For one thing, if the pasture has been grazed in the past year, there may well not be enough grass left to carry a fire (1,500 lbs/acre is about the minimum amount). Also, planning for the post-burn period is required, for there may be times right after the burn when a brief period of grazing is beneficial, but after that the burned area should probably be rested for one growing season. Finally, you need help to conduct a burn, and the best way to get experienced help is to

A good illustration of the benefits of prescribed burning. The prescribed fire killed the cedar in the center, the shin oaks on the left are resprouting from the roots, and the little bluestem in the background was not permanently harmed.

help others first. The best way to do that is to join the nearest Prescribed Burn Association to meet people who have experience and equipment and who will welcome your help and be willing to help you. Your local NRCS, TPWD, or TALE (Texas AgriLife Extension Service, formerly Texas Cooperative Extension Service) office can get you in contact with the nearest Prescribed Burn Association.

My main point about the prescribed burn is that it is not something to be afraid of, but it is a complicated subject that requires some time and effort to master. There is little question that almost every range can be improved by an occasional (every five to seven years) burn. It is how this country evolved with plant species adapted to periodic fire. Done properly, it does not hurt any mature hardwood trees or even mature cedars, but it does almost always lead to a more productive range from the standpoint of grass, forbs, and browse.

16

Protecting Your Home from a Wildfire

IN CHAPTER 15 we consider prescribed fires set intentionally to help manage a native pasture. But most of us also live in the country, and therefore a wildfire can be a threat to our homes and even our lives. A wildfire can occur under any conditions, but small, accidental fires are much more likely to become serious wildfires under conditions very different from those under which prescribed fires are conducted. Wildfires that occur under very dry (less than 20 percent humidity) and very windy (greater than 20 miles/hr) conditions, both of which contribute to the greatly increased intensity of a wildfire, are very difficult to control. Under these conditions grass fires can turn into crown fires that engulf mature cedars, jump from one cedar tree to another, and generate firebrands (pieces of burning limbs carried by the wind and rising air currents generated by the fire itself) capable of jumping hundreds of feet and setting new fires ahead of the main blaze.

Regardless of the type of fire, the cause, or the intensity, any combustible material in the path is at risk, and that includes our houses, garages, and other buildings. Although this may not be a land management issue, it is nonetheless an important issue to anyone living in the country. One important fact not always appreciated by country dwellers is that fire department help takes much longer to get to your place than to a house in the city, so chances of a fire being extinguished quickly are much less. This fact places a much greater burden on homeowners to do whatever they can to protect their homes from ever being involved in a fire.

Basically, what one needs to think about is what would happen if a wildfire burning in the pasture around you approached your home. Unfortunately, many people tend to answer that question by saying they would run outside and wet down their house with a garden hose. That is absolutely the wrong thing to do and potentially deadly. Some things to consider: (1) You might not be home at the time. (2) If the fire is that close, you really should be evacuating. (3) A garden hose cannot put out enough water to get things wet enough to protect your home from a raging crown fire. (4) You almost certainly cannot imagine the intensity of the heat, smoke, and raining firebrands you will experience in the path of such a fire. Even grass fires can move incredibly fast.

Having established what you should not do, what can you do? The first thing to do is—now, before a fire threatens—slowly walk all around your house and other buildings and try to envision what would happen if the vegetation around the buildings and out to as much as thirty feet or so were to be burning vigorously. Maybe today is a wet day in May and everything is green, but think about what it will look like on a hot, dry August day or a dry windy winter day. If you can see how the burning grass, shrubs, and trees could catch any part of your house, especially in the areas of the eaves or soffits where flames could get into the attic, then you have a potential problem. Now think about whether the burning vegetation could catch something else, such as a wooden fence or gate, a pile of firewood, flammable material under a deck, a propane tank, any of which could then catch some part of a building. If you can see how any of these types of things could act as a "ladder" to conduct the fire from the burning vegetation to your home, then you have a potential problem.

None of the above is an absolute. If the only burning vegetation that could get to your home is short, mowed, and watered lawn

A cedar brush pile was burned on an adjacent property, but it was not properly extinguished. The fire rekindled when the wind shifted, causing a grass fire that was luckily stopped just as the flames reached a neighbor's home.

grass, that is much less of a concern than if it is tall, unmowed, native grass or even worse pampas grass. If the only trees that could catch your house are live oaks or other hardwoods, then that is not nearly as serious a threat as if they are cedars. If your house is all stone with fiber cement trim and soffits, a metal roof, and double-pane windows, then your house is not nearly as vulnerable as if it has wood siding, single-pane windows, and cedar shake roof shingles. If the vegetation next to your house could be a threat, but it is protected from a wildfire by being inside a paved or gravel driveway so that a wildfire is less likely to ignite the vegetation near the house, then that situation is also less of a concern than one

Even the smallest grass fire could catch the pile of debris against this wood-sided building with a cedar shake roof. This is not the way to protect your home from a wildfire.

where there is a continuous connection between the native grasses and cedars in the pasture and the plants under your windows.

Now that you have taken that walk with an eye to the potential fire danger, many of the things you can do to help mitigate the threat will be obvious. If there are large, evergreen, especially flammable shrubs such as agarita, yaupon, cedar, or rosemary adjacent to your house, consider removing them, reducing their size, or making sure a wildfire is unlikely to get to these shrubs. This is just one example of many things you can do to better prepare your house to withstand a wildfire. For more ideas and information on this topic, go to the Texas Forest Service website, http://texasforestservice.tamu.edu.

17

Restoration

L AND restoration can mean different things to different people, but often it means replacing something that is growing in an area with something else or getting something to grow where nothing is currently growing. It may mean replacing non-native grasses, forbs, or woody plants with natives. It may mean replacing nonedible or noxious plants with good, edible, nutritious ones. Or, unfortunately, it may mean replacing perfectly healthy native vegetation with exotics. All of these things sound simple in concept, but accomplishing them is frequently much more difficult than it sounds.

Before beginning any restoration, it is important to take a hard look at the current conditions and ask a very important question: Why is xyz growing here and not abc? If you want abc but instead you have xyz, you need to know why that is. Often, what is growing in a particular place has been determined by the previous land management practices, in combination with the characteristics of the location (soil type and depth, drainage, slope, orientation, etc.). If that is the case, then simply planting abc seeds will not likely result in the replacement of xyz with abc unless something about the management practices also changes. This is particularly true of overgrazed areas; what is and is not growing there is likely the result of the overgrazing, and long term it is unlikely to change unless the grazing practices change.

The term restoration implies a disturbance from previous conditions or the ideal condition, and it may also imply a belief that this disturbance is temporary. People encounter many different situations that they wish to "restore." Here is a partial list:

1. Ground disturbance has occurred because of construction or heavy equipment, and the vegetation has been lost or damaged.

2. Cedar clearing has left bare spots as well as some ground disturbance.

3. The ground is bare around a stock tank, pond, or streambank where cattle go to water.

4. The pasture in general has too many short grasses, annuals, and unpalatable grasses and few if any of the better-quality plants.

5. A steep rocky hillside has very little vegetation and is eroding.

6. Areas where large brush piles were burned are now largely bare ground and a few thistles.

7. You want to convert an old cultivated field into native pasture again.

8. You have large areas of thick KR bluestem in the pasture and want to replace it with native grasses.

9. You want to turn the area in front of your house into a field of bluebonnets, Indian paintbrush, and gaillardia.

Each of these situations requires a different kind of approach. The first thought for most folks is to go out and buy some seed and reseed the areas of concern. Unfortunately, this can have a high probability of failure, partially because people often do not do a good job of planting the seed, partially because whatever seedlings do come up are not protected long enough for them to become established, and partially because of poor weather conditions, such as drought.

So, before we discuss how to handle each of the above problems,

This "restoration" was accomplished by fencing grazers and browsers out of the other side of the fence. The result is dramatic. Within three years the blackjack oak limbs had grown down to the ground, flame-leaf sumac and blackjack had sprouted and reached over six feet in height, and the seeded switchgrass became thoroughly established.

we need to discuss how to plant seed and maintain young seedlings—specifically, seed for native grasses and forbs on native rangeland, not cultivated crops. The first requirement is that the seeds make good contact with the soil but not be buried deeply. These seeds are very small and can grow only tiny roots and leaf shoots before they require moisture from the ground and sunlight. Any operation that either buries the seeds too deeply (more than 1/4 inch) or leaves some seeds caught in the leaf litter or barely touching the ground is wasting expensive seeds. Then, fresh seedlings need protection so that they can grow deep roots to become well established before they are grazed. And though in chapter 11 I mentioned that white-tailed deer can't eat much grass, it turns out that fresh new shoots have very little lignin and are quite edible, so even deer can and will eat these small shoots. Finally, if you were to buy a bag of lawn grass seed and read the directions, you would likely find the suggestion that the seed and seedlings be watered lightly twice a day, conditions you are unlikely to have on your range after you plant the seed, and the one thing you have no control over. So you can see why this whole process is considered risky.

Fortunately, native grass seed can remain viable for a long time, and if you are unsuccessful in getting germination the first year it can still happen the next or even two or three years thereafter. Also, fortunately, even a light seeding likely spreads many more seeds than you actually need to get new plants established. So, having warned about how risky reseeding is, I should also say that it can be and is done quite successfully. But it is also true that it is not always the first thing to try in all circumstances, as discussed below.

One final point about seeding: native plant seeds are expensive, so it is important that you use only seed that is adapted to our area. Little bluestem grows from South Texas to Canada, but the seed from outside our area is not adapted to our soils and climate, so it is important to know where your seed was grown. It is also true that, in general, you will have better luck with native seed mixtures, because even if the particular location where you want to plant is not ideal for one species, it probably is for other species in the mix. Also, instead of planting all of your seeds at one time, if you plant some at one time and the rest at another time, you double the chances that the weather will be in your favor. The catalog published by Native American Seed in Junction contains a lot of useful information about how, what, and when to plant.

Now, on to the problem conditions listed above:

1. Ground disturbance from construction. Assuming that the undisturbed areas next to the damaged areas have desirable native

vegetation, simply leaving the area for nature to restore usually works, but it may take several years. You also need to know that the natural plant succession in the Hill Country is for forbs to be the first plants to come up in bare soil, and frequently these include undesirable forbs (thistles, grass burs, etc.), but eventually other forbs and grasses usually take over. On the other hand, especially if the area is near enough to your house to be watered occasionally, planting native seed mixtures will restore the area much faster.

2. Cedar clearing. The best thing to do here is look around the areas that were not disturbed. If those areas have desirable vegetation, then almost certainly the seed bank in the whole area is sufficient to reseed the bare and disturbed areas without you doing anything. On the other hand, if the area cleared was a dense cedar brake with little or no grass between the bushes, then the seed bank may not be very good and you can expect to wait a long time before the area is revegetated. If you have a large area to restore, consult the local NRCS or Native American Seed people for ideas.

3. Bare ground around water source. Here there is no use doing anything unless the animals can be fenced out of the area. Once the animals are fenced out, the area will probably recover if just left alone (there should be lots of seed in the manure).

4. Too many undesirable grasses. In most cases, all that really needs to be done is to rest the pasture for a year or two and to then limit the stocking rate to well below the carrying capacity for several years. This should allow the decreaser (desirable) grasses to begin to come back and crowd out the less desirable species.

5. Steep rocky hillside. Much of the soil has probably eroded away from this hillside, which is why there is so little vegetation. Attempts to plant seed in areas such as this are likely to cause more erosion. If the area is being grazed, resting the pasture may improve the vegetation. Otherwise, consult NRCS agents.

6. Bare ground from burn piles. There have been two theories about why such areas take so long to recover: the burning killed

the microorganisms in the soil, or the burning killed all the seed in the soil. Some recent research indicates that it may be lack of seed, and that seeding will speed up the recovery.

7. Converting an old cultivated field to native pasture. This is a situation where plowing under the old crop, preparing the field, then seeding mechanically may be the best method. The old growth may need to be treated with herbicide either before or after plowing under to prevent aggressive grasses (e.g., bermudagrass) from competing with the native mixture you are planting.

8. Replacing KR bluestem with native grasses. This may well be a situation where the cure is worse than the disease. The only sure way to get rid of the KR is a large-scale herbicide treatment, which will certainly also kill a lot of native vegetation. Furthermore, the KR has produced huge quantities of seed that may compete with the native seeds if reseeding is used. There is some evidence that resting pastures and allowing the tall native grasses to increase tend to reduce the amount of KR or keep it from spreading. There is also some research being conducted with specifically timed prescribed burns, but this remains one of the more difficult "restorations" to accomplish.

9. Add bluebonnets, Indian paintbrush, and gaillardia. This is a common desire of many new landowners. Unfortunately, the ecology of the Hill Country is working against them. First, these annual forbs naturally germinate and grow in disturbed, bare soil, not in heavy native grasslands. If an area is purposefully graded or plowed to make such bare soil, then the intended forbs may have to compete with undesirable ones whose seeds are also present in the soil. But the main obstacle to accomplishing this dream is that these annual forbs are among the white-tailed deer's favorite food. Consequently, in most areas of the Hill Country the deer population will eat these flowers down to nothing. The alternatives are to fence out the deer or limit the wildflowers to those not eaten by deer, such as Mexican hat, mealy blue sage, and prairie verbena. Otherwise, your efforts will almost certainly be disappointing.

18

Managing for Songbirds

MANY people enjoy birds. Even people who have no gardens, flowers, or other plants growing around their house, spend little time outdoors, and have little interest in nature frequently hang bird feeders from trees. And, increasingly, many landowners have converted their agricultural tax valuation to a wildlife tax valuation and are actively managing their land for songbirds.

How do you encourage birds to hang around your place? The first thing is to try to look at your place as a bird would. A bird needs several things: food and water are obvious; cover from predators is important for some species; shelter from winter winds and the heat of the day is important for others. Some look for a place to build a nest if they nest in the Hill Country, and some species need space.

Although the Hill Country is a great birding area, the average Hill Country property may well be lacking in at least one of these requirements, at least as seen through the eyes of one of our avian friends. To attract birds to your property, you need to manage your property in a way that produces the kind of habitat that satisfies as many of these requirements as possible. No one property will likely attract every species that is present here, but you can certainly improve the number with relatively little effort.

Food is probably the first thing people think of when trying to entice birds or other animals to their property (although water may actually be more important), and buying a feeder and some seed is pretty easy to do. Sunflower seeds probably attract more species than anything else (black oil is the most nutritious), although if you want goldfinches you will have better luck with thistle seeds and feeders. The mixed wild bird food has some seeds that may go to waste or may attract house sparrows and white-winged doves. If you are going to be upset that squirrels partake of your offerings as well, then you have to be careful about the type of feeder you have and where and how you hang it. One of the main things to think about is consistency; if you maintain your feeders regularly for several months and then quit, the birds that have relied on your food will have to find another source or suffer.

Not all of the seeds your seed eaters want have to come from the store. Plant seed-producing perennials such as tall or prairie goldenrod and sunflowers and you will find finches and many smaller birds flocking to them. Other birds may like Mexican hat, any of the crotons (also known as doveweed), Engelmann daisy, and most of the grama grasses.

Of course, there are also the hummingbird feeders, and because of these bird's relative tameness and antics, many people find them the most entertaining of all. The sugar water feeders (just use one part pure sugar and four parts water; no red dye!) can be hung close to your house—these birds are not that afraid of you and will actually feed just inches from you. Growing native flowering plants in your yard also attracts these little creatures and gives them some variety of nectar. Some examples of native plants that attract hummers are red yucca, Turk's cap, flame acanthus, coral honeysuckle, trumpet creeper, salvias (autumn sage, tropical sage, mountain sage, and big red sage), standing cypress, penstemons, and columbines.

The problem with all of these plants is that they do nothing for all of the insect eaters, raptors, or carrion eaters. There may not be

Providing feed for songbirds not only helps the birds but provides many hours of entertainment and enjoyment.

much you are willing to do with the last two groups, but you can help out the insect eaters. For most of them, when the insect populations decline, they can switch to berries or in some cases other fruit. Fortunately, Mother Nature has provided a good source of berries, timed just right to correspond to the decline in insects in the late fall/winter. By planting any of the berry-producing trees or shrubs, you can make life easier for the insect eaters. Trees such as hackberry, juniper (cedar), Western soapberry, and escarpment black cherry are good choices. So are such shrubs as American beautyberry, agarita, rough-leaf dogwood, prairie flame-leaf sumac, skunkbush sumac (fragrant sumac), Mexican plum, Carolina buckthorn, possumhaw, and yaupon as well as wild grape and coral honeysuckle vines.

The basic food idea is that the best habitats have as many natural food sources as possible, in terms of both quantity and number of species. In addition, diverse vegetation attracts the most species of insects, which also attracts a wider range of birds.

Certainly water has to rank on a par with food as a necessity for birds. Birds can fly long distances for water, but they certainly take advantage of closer sources, so if you make one available you will have more birds more of the time. Providing water can be as simple as a shallow pan or "bird bath" you keep filled, or as elaborate as a fountain or pond. One word of caution: a deep bucket or tub is not a good type of water source, because if a bird gets into it and gets wet it may not be able to get out easily; the best water sources are shallow. Running water that makes a noise is also attractive to birds. For years we simply had a few shallow dishes that we filled with water, placed around in areas near the bird feeders. When we then installed a small recirculating stream and waterfall with shallow depressions in the rocks, we began to see many more birds use the water and even a few different species we had not seen before. The hummingbirds and summer tanagers love the new water feature but never used the "bird bath" features.

Birds need cover to hide from predators, most commonly raptors. If they are easily seen from the air because they are not under any vegetation, then they are especially vulnerable. Quail in par-

It is not necessary to buy fancy bird feeders. The birds seem to like the sun-flower seeds sprinkled on this old log just fine.

ticular have specific requirements: shrubs to hide from predators and tall grass to hide their nests, but enough space between tall grass clumps to roam and search for food. These rather strict requirements may be one reason quail are declining in Texas, since there are fewer places that fulfill them.

In the heat of the day birds need shade, which to them means trees and large shrubs. If your area is largely free of trees and shrubs, then your ability to attract birds will be limited in the summer. Likewise, birds need protection from cold winter winds, and dense evergreen shrubs (e.g., cedar, Texas mountain laurel) seem to be the best type of protection.

Some birds have rather strict nesting requirements. The black-capped vireo, for instance, wants to build its nest about three feet off the ground. If your land has a distinct browse line up to about five feet with virtually nothing growing between the browse line and the grass, then your area is not as attractive to them or many of the other birds, like cardinals, that typically build nests below six feet.

Some birds like nest boxes or birdhouses, but again a diversity of sizes (especially the size of the hole) and locations is required to have one suitable to every species. There are many books as well as information on the Web that offer instructions for building and positioning birdhouses. Some birds (bluebirds and purple martins) need some space around the nest box; others like nest boxes that are concealed.

Some birds, such as quail, meadowlarks, turkey, and purple martins, like large areas of open space. Some, like the scissortail flycatchers and the kestrel, prefer high open perches. Other birds seem to thrive in the smallest spaces in the densest woodlands, and it is obvious that no one habitat appeals to them all.

The key concept for managing songbirds is diversity: diversity in species, diversity in size and density of vegetation, diversity in food sources. The ideal habitat has short grasses and forbs in places, taller perennials in other places, shrubs of different size with different types of seeds or berries, and a variety of species of trees, including some deciduous and some evergreen. Most people do not have all of that growing on their property, so to achieve that goal some planting and gardening needs to be done. But the good news is that you can start wherever you are, make any changes at your own pace, and do as much as you wish. You may notice that I just described the ideal habitat for birds as a healthy mixture of different species of grasses, forbs, shrubs, and trees. I hope you recognize that description as the same one I give throughout this book for a healthy, diverse, sustainable habitat. So, if you build a good habitat, they will come.

One final note, and one some folks will not like to hear. It is best illustrated by a story from a friend of mine. He met a woman who wanted him to look over her property and identify some plants and talk about what she should do to attract more birds. As they walked around her property, she repeated her desire to attract more birds. Then, as they finished up their walk they went around to the back porch, where my friend saw a half dozen cats lounging in the shade. Cats that are running loose outside—whether you call them pets, house cats that you let out occasionally, "outside" cats, or feral cats—are a real menace to songbirds, and you will never have the numbers or diversity of birds around your place if you have cats outside. Also, if you do have cats outside, please position all the feeders and water sources only in open areas where the birds can easily see a cat approaching them, not where a cat can hide behind something and pounce on the birds.

Endangered Species

This seems to be the place to mention two endangered species: the golden-cheeked warbler and the black-capped vireo. They are both on the federal endangered species list, and they both nest in the Hill Country. In fact, the only place on Earth that the golden-cheeked warbler nests is in Texas.

There is perhaps as much misinformation about the habitat requirements of the golden-cheeked warbler as there is about anything else in the Hill Country. Ideal habitat for this little bird is a mixed forest of mature cedar and hardwoods with a large area of fairly dense canopy along the side of a canyon. The reason for mature cedar is that the bird needs the peeling strips of bark from a mature cedar to build its nest. But it doesn't necessarily build its nest in the cedar, preferring usually a large Spanish oak or other hardwood. A cedar brake of fifteen- to twenty-foot trees with few hardwoods is not good habitat for the golden-cheeked warbler, so cutting these trees is not destroying their habitat. A handful of cedars and oaks on a half acre is not likely to appeal to them either. But if you have the kind of habitat described above, you might want to consult a bird expert before cutting down the big old cedar trees. Management guidelines for the golden-cheeked warbler can be found on the Texas Parks and Wildlife website (see chapter 23).

The black-capped vireo has a very different kind of habitat requirement. It likes thick brush low to the ground where it can build its nest about three feet off the ground. The density of low-growing shrubs seems to be the most important aspect; it doesn't need tall trees, and cedar seems to hold no attraction for this little bird. The main threats to the welfare of the black-capped vireo are the brown-headed cowbird parasitizing its nest (see below) and browsing white-tailed deer, goats, and exotics literally eating away at its nest sites. This is just another example where too many of one species threatens another species, and where a knowledge of the habitat requirements allows a landowner to help protect the threatened species.

The brown-headed cowbird evolved to migrate with the bison herds, thus moving constantly. Because they were never in one place very long, they didn't have time to make nests, incubate eggs, and raise the young, so they learned to lay their eggs in other bird's nests. Fortunately for the cowbird, but unfortunately for other birds, cowbird eggs hatch about two days faster than other bird's, so the cowbird young outcompete the host birds' young for food, and the host birds' young frequently do not survive. Fortunately, many birds are able to make two nests in a year, and the bison-following cowbirds were likely gone before the next nesting period, so the host bird could still raise young.

But now that the bison are no longer migrating across the plains, the cowbirds flock to herds of cattle, which do not migrate, so the cowbirds don't migrate either and are present year round to parasitize other birds' nests and thus cause even more problems for them, the black-capped vireo among them. Research has shown that, where cowbird populations have been reduced by trapping, the populations of other songbirds have increased.

The Texas Parks and Wildlife Department has a program to teach landowners how to trap cowbirds. Anyone interested in the program can contact their local TPWD office.

19

Native Grasses

GRASSES are arguably the most important component of the landscape. The ideal, sustainable, healthy habitat in the Hill Country is a mixture of trees, forbs, and grasses. Remove any one of these classes and the diversity and quality of the habitat decrease, along with the number of animal species that can be sustained. But of the three types of vegetation, grasses are the best at keeping the soil in place and in good condition, which is the prerequisite for a long-term sustainable habitat.

Biologists, naturalists, and historians all talk about the tall grass prairie that early settlers encountered in the midsection of this country in the 1800s. The physical characteristics of these prairies, which stretched from Texas to Canada, determined in large part the speed with which these areas were settled and the structure of these settlements. The most common grasses found in almost all of the tall grass prairies were what are referred to as the "Big Four": little bluestem, big bluestem, yellow indiangrass, and switchgrass. All four of these grasses produce large amounts of very edible, nutritious vegetation. Little bluestem can produce seed heads in excess of five feet high, and all of the others commonly produce seed heads in excess of six feet—the early accounts of grass "as high as a stirrup" or as "high as a saddle horn" were not exaggerations.

Big bluestem growing in a remnant of a tall grass prairie just a little north of what would be considered the Hill Country.

Virtually all of the original tall grass prairie is gone now, but a visit to any of the few remnants is an eye-opening experience. You can quickly lose sight of your companions as you have literally to push your way through thick grass cover taller than your head, and your feet never touch the ground.

Though remnants of original prairies are rare, the Big Four grasses, along with many other native species, are still with us. Some, like little bluestem, are quite common; others, like big bluestem, are relatively uncommon. But there are many ranches in the Hill Country that support good stands of at least three of the Big Four.

The practices of those ranches that do have such good grasses have taught us all a lot about range management.

The following is a brief discussion of several of the more common native grasses in the Hill Country. To identify these grasses, I recommend using one or two of the grass books listed in the Bibliography.

Little bluestem (*Schizachyrium scoparium*) once made up close to half of the grass plants in the Hill Country, and it is still one of the more common species. It is one of the dominant grasses seen in all well-maintained ranches, and it is one of the main grasses to reestablish itself when abused land is allowed to rest and recover. It is highly palatable for the first half of the year but begins to be less so in July as it becomes more lignified. The round clump of leaves makes for excellent quail habitat as well as cover for other ground-nesting birds. Frank Gould, author of *Common Texas Grasses*, notes that "the presence of vigorous stands of little bluestem on a range is a general indication of good land and good management." In the fall it is easily identified even from a distance, its multiple erect stems turning a rusty brown in contrast to the lighter color of most other grasses.

Yellow indiangrass (*Sorghastrum nutans*), another of the Big Four, shows its family resemblance to cultivated sorghums with its tall, erect stems and large spreading seed heads containing large seeds. It is probably the second most commonly distributed of the Big Four and is a good indicator grass for well-managed properties. It can also be used in landscaping to replace the nonnative pampas grass.

Switchgrass (*Panicum virgatum*) is most often found in moist or semimoist areas and can produce large amounts of forage from a single plant. It probably most resembles nonnative johnsongrass, although it lacks the white midrib and fungal patches seen on johnsongrass, and its seed head is finer with smaller seeds. Also, I have noted that, if you hold a leaf blade between the thumb and forefinger and slowly pull the blade from the base toward the tip, somewhere in the last third of the leaf you will feel an "imperfection" in the leaf that may or may not be visible. I have never seen this feature described in any book, but I have never seen it to fail and have not observed it in any other grass.

Big bluestem (*Andropogon gerardii*) is the least common of the Big Four grasses today, at least in the Hill Country, possibly because it is a real favorite with grazers and is easily grazed out of today's pastures, and it may be less drought tolerant. It can be recognized by its distinctive seed head with two to five branches at the top of the stem, but most often it has three branches spread out in a pattern that suggests a turkey foot, and in fact the grass is sometimes called the turkey foot grass. It is frequently missing in fields where the other three Big Four are abundant.

Some of the other palatable, good-quality grasses that are also indicators of well-managed ranches include sideoats grama (*Boute-loua curtipendula*), the state grass of Texas; eastern gamagrass (*Tripsacum dactyloides*), another real favorite of grazers that is usually found near water; Canada wildrye; cane bluestem; silver bluestem; green sprangletop; plains lovegrass; and Texas winter-grass, also known as speargrass. Most all well-managed pastures have at least half of these species.

Grasses most likely to be seen near water include bushy bluestem, Lindheimer muhly, switchgrass, eastern gamagrass, seep muhly, and vaseygrass (nonnative). Bushy bluestem is almost al-ways associated with a water source, either obvious surface water or a seep. Most of the others can be found at times somewhat re-moved from obvious wet areas.

Cool-season grasses that green up early in the spring and pro-vide green forage as early as February or March include rescuegrass (introduced), downy brome-(introduced), Canada and Virginia wild-rye, Texas wintergrass, and Scribner's dichanthelium. Switchgrass is considered by some, but not most, authorities to be a cool-season grass, since it puts out a lot of growth before June 1, but it doesn't start nearly as early as the other cool-season grasses.

Short grasses with poor forage quality and quantity that are good indicators of overgrazed fields include Texas grama, red

Switchgrass, growing in native Hill Country soil, dwarfs the author in a wet year.

grama, sixweeks grama, needle grama, purple threeawn, most all other threeawns, hairy tridens, windmillgrass, and tumblegrass. A predominance of these grasses, even if small amounts of better grasses are also present, is probably an indication of past or present overgrazing or, in some cases, very thin, rocky soil.

Almost all of the native grasses are what are called bunch grasses because they grow in clumps rather than spread out into a ground-covering lawnlike turf. The exceptions are three native grasses that also have stolons (runners on top of the ground that allow the grass to spread like bermudagrass): curly mesquite, buffalograss, and vine mesquite.

There are a number of introduced grasses, some of which can become invasive and threaten native pastures. Among the more common such grasses seen in the Hill Country are bermudagrass, johnsongrass, dallisgrass, KR bluestem, Kleburg bluestem, kleingrass, Old World bluestem, Willmann lovegrass, weeping lovegrass, stinkgrass, ryegrass, and Japanese brome. Of these, bermuda, in all its various cultivars, johnsongrass, kleingrass, and ryegrass are purposely cultivated as "improved" pastures for livestock. The Texas Department of Transportation has, unfortunately, planted KR bluestem all over the state along roadsides to prevent erosion. The result has been the widespread invasion of the grass into many Hill Country pastures, especially those that have been somewhat overgrazed and are thus lacking in a good diversity of native species. KR is less desirable than most native grasses from an animal forage or wildlife standpoint, but its eradication is quite difficult.

I have tried throughout this book to use the common Hill Country names for all of our native plants, rather than the less familiar, unpronounceable scientific names. It is, however, undeniable that people in different parts of the state have different common names for the same species as well as the same name for different species. So, in order not to be misunderstood about exactly which species I am referring to, I include the scientific names for all species mentioned in appendix 2 of this book. So please, if there is any concern about exactly which species is meant, refer to the tables in the back and note the unambiguous scientific name.

20

Native Woody Plants

MOST people's mental vision of the Hill Country includes some combination of hills and trees; my vision certainly does. It may include grass and it may include wildflowers, but it certainly includes trees. If we had been around 150-plus years ago, there would not have been so many trees, but the Hill Country was never the "treeless plains" that settlers found in areas farther north.

Many people choose which property to buy based largely on the trees growing there. Lots of folks choose house sites based on the proximity to favorite trees. But trees are important for more than just the aesthetic value we humans place on them. Trees add considerably to the diversity of habitat, providing food, shelter, nest sites, and cover for many bird species and other animals. They also provide shade for vegetation growing under them, a fact made obvious at times when the grass is green under trees while it has gone dormant out in the open.

Many if not most species of native trees and shrubs in the Hill Country are declining in number almost totally because of the overpopulation of deer consuming all the small replacement trees and shrubs within their reach. If for no other reason than that, one can make an argument for preserving and protecting as many

The Hill Country live oak. Where would we be without it?

native trees as possible—the obvious exception being conditions where cedar or any other species expands to form thickets that crowd out all other species. Under these conditions, biodiversity is reduced and these areas should be managed to limit such unrestrained growth, as discussed in chapter 9.

There are numerous reference books to aid one in identification of woody plants in the area, many of which are listed in the Bibliography. Unfortunately, most of these books are organized alphabetically by botanical family, which is pretty much useless in terms of plant identification by the novice, who is usually left with thumbing through the pictures looking for a match. If you are primarily interested in Hill Country plants and expect to buy only one book, Jan Wrede's *Trees, Shrubs and Vines of the Texas Hill Country* will probably be the most useful. The alternative is to have a biologist from one of the state agencies, or a Texas Master Naturalist, or some other knowledgeable person walk your property with you and help you identify what you have.

The following is a brief discussion of the more common or important Hill Country trees and shrubs. As with the grasses, I use the most common Hill Country name for these woody species, but you can refer to appendix 2 for the more precise scientific name. This is especially important if you are planning to buy trees at nurseries to plant on your place. Always read the label for the scientific name before buying, for erroneous or misleading common names are frequently used by nurseries, and you can wind up planting something far different from what you intended.

Large Trees

Cedar, or more correctly Ashe juniper, is discussed in chapter 9.

Live oak is probably the second most common tree in the Hill Country, after cedar. The more common variety is the plateau live oak (*Quercus fusiformis*), but the more eastern variety (*Q. virginiana*) or a hybrid of the two grow here as well. On many properties, especially well away from permanent water and on slopes or thin rocky soils, live oak and cedar are the only major trees to be found. Because of live oak's abundance and evergreen nature, its leaves are a major part of the diet of the white-tailed deer in overpopulated areas.

Spanish oak, or Texas red oak, is a member of the Red Oak family. They are common on rocky hillsides. They provide light green foliage in early spring and usually good red fall color as well. They tend to be somewhat short-lived because they are brittle , which causes them to succumb frequently to wind damage. Because they are such a favorite of deer, their numbers are declining.

Blackjack oak is the other Red Oak family member of the Hill

Country. They have a darker bark than most other trees and tend to be slightly smaller than live oaks and Spanish oaks.

Post oak is characterized by a lighter-colored bark than most other trees; typically the trunks are unusually straight and vertical. Post oaks are frequently found growing along with blackjack oaks in redland soils. As members of the White Oak family, they are not very susceptible to oak wilt.

Cedar elm is a strong, robust, relatively disease free tree with small, rough leaves, which tend to turn a light yellow in the fall.

Escarpment black cherry has small, white blooms in spring, but they produce only very small fruit, favored by birds. The leaves turn yellow in the fall. Small trees or limbs (less than three inches) tend to have whitish flaky bark with conspicuous rings, but mature trees have very dark bark.

Lacey oak, also called blue oak, is a slightly smaller oak than the other large oaks, and its leaves have a bluish cast that distinguishes it from other trees. Lacey oaks can grow on rocky slopes and thin soil but also in deeper soils in canyons and draws.

Chinkapin oak, also spelled "chinquapin," is a large tree of deeper soils along creeks. The oblong leaves with very coarse-tooth edges are unique among oaks in this area.

Sycamore is very large, fast-growing tree usually found along streams, where it produces many saplings from seeds that lodge among the rocks of streambeds.

Bald cypress, another fast-growing, very large tree, is virtually always found near water. It is the only large tree with "needles" rather than leaves in the Hill Country.

Native pecan is almost always found in deep soil in creek bottoms or along streams and rivers. These trees can become very large, and the nuts provide food for several species of wildlife. They can be distinguished from walnut trees by their multipart husks, larger leaves and leaflets, and the twigs not showing chambered pith when sliced open lengthwise.

Walnuts in most of the Hill Country are Texas walnuts (aka little walnuts), which are usually less than thirty feet tall and have small (less than an inch) nuts. They are also frequently found along streambeds but can tolerate much thinner, poorer soils and less moisture than pecans require. In the eastern part of the Hill Country their range overlaps that of the eastern black walnut and in the west that of the Arizona walnut. Walnut trees tend to lose their leaves in the fall earlier than most other trees, sometimes in early September.

Bigtooth maple is found natively in the Hill Country only in creeks and canyons of Bandera, Kendall, Real, and Uvalde counties. They are the main attraction in Lost Maples State Park for folks seeking fall color. They seem to grow well in most landscapes.

Hackberry for some reason has a bad reputation among many folks, but in fact this is probably the native tree used by more species of wildlife than any other tree in Texas.

Small Trees and Shrubs

The number of relatively common small trees and shrubs in the Hill Country is too great to discuss each one in any detail, and they are all discussed in most of the books on woody plants listed in the Bibliography. Collectively, all of these species contribute significantly to the diversity and quality of habitat in the Hill Country, and any well-managed ranch has many of them.

The species of this group that would most often be considered small trees include Eve's necklace, goldenball leadtree, gum bumelia, rusty blackhaw viburnum, Texas madrone, Texas persimmon, Texas redbud, and white shin oak (aka Bigelow oak).

The most common multitrunk shrubs seen in the Hill Country include agarita, American beautyberry, buttonbush, Carolina buckthorn, elbowbush, evergreen sumac, prairie flameleaf sumac, fragrant sumac, kidneywood, Mexican buckeye, Mexican silktassel, mesquite, fragrant mimosa, possumhaw, rough-leaf dogwood, Texas mountain laurel, toothache tree, yaupon, yellow buckeye, and willow baccharis (invasive). Note that my distinction between a small tree and a shrub here is somewhat arbitrary; some individual "small trees" can be multitrunk shrubs, and vice versa.

Common vines include greenbriar, grape, dewberry, Virginia creeper, trumpet creeper, and poison ivy.

The most common cacti include two species of prickly pear (*Opuntia engelmannii, O. macrorhiza*), lace cactus, claret cup, and horse crippler. The most common succulents are twist-leaf yucca, Buckley yucca, nolina, and sotol.

21

Miscellaneous Topics

Ball Moss and Lichen

MOST everyone from East Texas to the Carolinas is familiar with Spanish moss. Ball moss is a cousin to Spanish moss; they are both epiphytes, which means they get their nourishment from the air and rainfall. They are not attached to the vascular structure of trees, they derive no nourishment from the trees, and therefore they do not cause any damage to the limbs. Since ball moss is frequently, but not always, seen on dead lower limbs of live oaks, many folks assume the ball moss killed the limbs. The cause-and-effect assumption here is just wrong. The lower limbs of live oaks frequently die on healthy trees, largely because of lack of sunlight on the leaves, and this happens whether or not ball moss is around. It just turns out that ball moss likes the environment of the lower limbs (shaded, cooler, higher humidity), although it can also be seen growing on telephone and power lines in full sun.

Some people have tried to rid their trees of ball moss by spraying various chemicals on it. The problem they then discover is that they have dead ball moss attached to the limbs, which was not exactly what they had in mind.

Lichen is a symbiotic complex of algae and fungi that attaches to solid surfaces such as rocks and tree limbs. It likewise does not obtain nourishment from the tree, does not harm the tree, and is not a symptom of any disease. Lichen and ball moss are simply two other native plants that grow in the Hill Country. Some birds use them as nesting materials.

How to Plant a Tree

Most of us think we know how to plant a tree, but in fact the majority of newly planted trees that die do so because of improper planting procedures. The following set of directions, by Robert Edmonson of the Texas Forest Service, provides excellent instructions for planting trees and shrubs.

TREE PLANTING • The 12 Step Program

1) Select an appropriate location for the tree.

Use a tree that will grow well under local environmental conditions and provide it with plenty of space to grow and mature. This includes both vertical and horizontal space for the canopy and plenty of room for root growth.

2) Dig the hole at least twice as wide as the root ball (wider is better).

Wide areas give roots a place to spread and grow. Dig the hole no deeper than the root ball to keep the tree from settling too deep and dig square holes to allow for root penetration out of the hole and into the surrounding soil.

3) Fill the hole with water and check the drainage.

If it takes longer than 24 hours to drain, select another site or another tree. A tree will die if its roots are underwater for long periods of time. Tree roots need air.

4) Prune the tree sparingly only if necessary.

Remove dead, broken, and diseased branches and crushed and girdling roots only. Removing even a small portion of the healthy canopy actually slows root growth and delays establishment. A thoughtfully selected tree requires no pruning.

5) Remove all foreign materials from the tree.

This includes wires, twine, cords, containers, tags and especially non-biodegradable bags. If planting a balled and burlapped tree, remove as much of the burlap as possible to allow for water infiltration into the bag and root penetration out of the bag.

6) Set the tree in the hole with the root collar flush or slightly above natural grade.

Planting too deep is a leading cause of mortality of newly planted trees. Do not pick the tree up by the trunk. Always handle by the container or root ball.

7) Gently backfill with the same soil that came out of the hole.

Create a natural environment, not an artificial one. Do not add soil amendments or fertilizer. Too much nitrogen will burn tender young roots, slowing growth and delaying establishment. Settle the soil with water. Tamping the soil causes compaction and damages roots.

8) Stake the tree only if necessary.

Consult a professional if staking is required. Stakes should not be left in place longer than 1 year.

9) Mulch the tree out to the drip line, 1–2 inches deep, and up to but not touching the trunk. Wood chips, pine bark, leaf litter, hay, etc. are great mulches. Mulch keeps soil temperature fluctuations to a minimum and increases soil moisture retention. Mulch also suppresses weed growth and organic mulch adds nutrients to the soil.

10) Water the tree for at least one year, preferably two.

A newly planted tree requires 6–8 gallons of water per diameter inch of trunk per week, less often in late fall or winter. A thorough soaking is much better than light, frequent watering.

11) Protect the tree from animals (this includes humans).

A wire-mesh cage at least 3 feet in diameter and 4 feet tall staked to the ground works miracles. Deer and livestock will eat your tree if it is not protected and weed-whackers will kill your tree in an instant.

12) Perform routine maintenance for at least two growing seasons.

This includes biannual weed control, yearly mulch replacement, weekly watering, and protection maintenance. Prune and fertilize only if necessary.

Selecting Trees and Shrubs for Planting

Different people have different lists of plants to recommend for planting in the Hill Country, and mine is certainly no better than others. I have tried to list in the following table only species that can usually be found at native plant nurseries and are reasonably easy to grow. That doesn't mean that a sycamore will grow on a rocky hilltop or that a Texas persimmon will grow in a seep, and it doesn't mean that anything will survive being eaten by deer.

While we are discussing planting trees, I should say that in general the best time to plant trees in the Hill Country is October or November. I know that many people grew up in other parts of the country thinking spring was the time to plant trees, but spring planting gives newly planted trees less time to grow roots into the new soil before the really hot summer period of stress arrives. This doesn't mean you can't plant trees in the spring, just that it is harder to keep them moist during the summer.

Blanco crabapple	Thorns, white blooms in spring
Buttonbush	Needs lots of water, ornate white spherical flowers
Carolina buckthorn	Shiny leaves and red to black berries, no thorns

Ball moss growing on a telephone wire, proving that it does not obtain nutrients from the trees it grows on (see page 150).

Cedar elm	Tough, sturdy shade tree
Chinkapin oak	Needs deep soil
Cypress, bald	Best near water, can survive elsewhere
Elbowbush	First bush to green up in the spring
Escarpment black cherry	Yellow in fall, tiny cherries for birds
Evergreen sumac	Evergreen, red berries in winter
Eve's necklace	Pink blooms in spring, bean pods constricted between beans
Flameleaf sumac	Birds love berries, red color in fall
Gum bumelia	Tiny white blooms in summer, thorns
Hackberry	Good for wildlife
Lacey oak	Blue-green leaves, can grow in thin soil
Live oak	Evergreen, grows almost anywhere
Mexican silktassel	Evergreen, not a deer favorite
Mimosa, fragrant	Pink blooms, delicate foliage, thorns
Mulberry, red	Birds love fruit
Pecan	Needs deep soil, nuts for wildlife
Possumhaw	Red berries last all winter
Rough-leaf dogwood	Makes thicket with root sprouts, good wildlife cover, berries
Shin oak	Small oak with flaking bark
Spanish oak	Early light green color, good red fall color, deer favorite
Sycamore	Needs to be near water
Texas ash	Fast-growing shade tree
Texas mountain laurel	Great, fragrant purple blooms, evergreen
Texas persimmon	All wildlife like fruit
Texas redbud	Pink flowers in spring
Toothache tree	Thorns
Wafer ash	Interesting blooms, seeds
Walnut	Small round nuts

Tree Pruning

The subject of pruning and painting oaks trees is reviewed in chapter 12 as it pertains to preventing oak wilt. The more general topic is the need to prune any mature native tree. Many people believe that trees must be pruned to keep them healthy. I have discussed this issue with several Texas Forest Service agents and asked specifically if there is any biological reason to prune a mature native tree. The answer I have received has always been "no." If you have a limb hanging over your house or car or driveway, then you probably have a good reason to prune it, but you are not doing it for the health of the tree. Even pruning dead limbs is unnecessary. All of our native trees evolved without humans to take care of them and have survived quite well. They do not need our help. Any pruning of mature trees is for our own benefit, be it to make room for us and our buildings and vehicles or for our view or purely for aesthetics. Pruning saplings to prevent crossed limbs or "V"-branching is a different issue, one that I am not discussing here.

Rainwater Harvesting and Water Supplies

According to Texas Environmental Profiles (www.Texasep.org), the population of the four counties of Bandera, Gillespie, Kendall, and Kerr in 2000 was 103,037, and the water use in those counties was 23,458 acre-feet. By 2020, these numbers were expected to increase to 168,683 people using 32,139 acre-feet of water. It is obvious that the number of people living in the Hill Country and surrounding cities will continue to grow for a long time, and this inevitably means the demand for water will also continue to grow. But the supply of water is assuredly not going to increase. Either the aquifers in the Hill Country are decreasing in volume with time or their level fluctuates with rainfall. Stream and spring flows are dependent upon rainfall amounts, which is not going to change (although some predictions for global warming predict a decrease in rainfall in this area). So the demand for water will increase, but the supply will not.

It doesn't take a rocket scientist to figure out that at some point, somewhere, the supply will be less than the demand. Better land management will do a better job of capturing the rainwater and storing it in the ground, where it can slowly seep out in springs and seeps and maintain constant river flows. But even though better land management is important, it will not prevent aquifers from being depleted and supplies exhausted by an ever-increasing population. Rainwater harvesting is one way individuals can ensure themselves of a supply of clean water in an environmentally friendly way.

The average thirty inches of annual rainfall in the Hill Country falling on a 2,500-square-foot roof yields 45,000 gallons of water, which is 123 gallons per day. This is more than enough for the average household practicing even very modest conservation. In addition, the water is essentially as pure and soft as distilled water. There is of course an initial cost to install such a system, but that is not necessarily more expensive than drilling a water well. Once installed, the maintenance is again comparable to or less than that required from a well system. I know at least a dozen people who have such a system; none of them have voiced any serious complaints, some really love the soft water, and none of them have ever run out of water. I highly recommend that everyone give serious consideration to such a system, especially if they are designing or building a new house, for which the installation is easier than in an existing home.

Understanding Rainfall

You may be asking yourself, What is there to understand about rainfall? Well, maybe more than you might think. As this is being written in early March 2008, I can look back over my records and see that in 2007 we had approximately fifty inches, which is about twenty inches more than average, or in other words an extremely wet year. From this you might conclude that things are in very good shape, moisture-wise. In fact, it is extremely dry, and we have been having many wildfires and red-flag warning days. Why? Because the last rain we had of over an inch was almost exactly six months ago, and since then we have received only slightly more than two inches total. From an agricultural standpoint, from the standpoint of moisture in the ground and in the vegetation, annual rainfall is almost meaningless. What is important is moisture in the ground. Very light rains (0.1 inch or less), even if we get them every day or two, put very little moisture in the ground because most of the moisture holds up in vegetation and evaporates. A four-inch rain doesn't put twice as much moisture in the ground as a two-inch rain, because a higher percentage of the larger rainfall runs off. If I could design a perfect rainfall system, we would have a one-inch rain every other Tuesday (a lot of Mondays are holidays), so we would get twenty-six inches in a year, or just over 85 percent of our average. But I assure you that we would have much healthier vegetation and more productive crops than we do now. It's not the total rainfall that is important, but how much falls in each rainfall event and how frequently we have rainfall. Both light rains and heavy downpours put a smaller percentage of the water into the ground than intermediate-size rain events. Most of a one-inch rain reaches the ground and soaks in without running off.

If you want to keep track of the condition of the vegetation on your land as well as the soil, you really need to keep records, and to do that you really need a rain gauge. You can't rely on how much rain fell at the local airport, because we all know that Hill Country rains are far from uniform. And unless you keep records of how much rain you got when, you won't be able to compare different times in terms of how productive they should have been. Besides, just the act of going out and checking the rain gauges and recording the data starts you in the habit of keeping a journal, and in the future you will be glad you did that.

Rainwater storage tanks. Capturing rainwater is likely becoming more and more of a necessity in the Hill Country as the population increases but the water supply does not.

Mowing

Many people have spent much of their lives living in the suburbs, where all of the neighbors are close by and can observe, sometimes critically, the condition of their yards. They all thus feel compelled to have a lawn and to keep it mowed. Nonnative lawn grasses (bermuda, St. Augustine, zoysia, etc.) are not a subject for this book. Basically, I don't recommend large lawns for folks living in the country, because they waste water and require frequent chemical treatments. If folks really want a lawn, I recommend one as small as possible and of native grasses, especially buffalo, blue grama, or curly mesquite. (Blue grama is not actually native to the Hill Country, but to regions north and west of here, but grows quite well here.) Buffalograss is available as seed or sod, blue grama as seed; curly mesquite will likely be available soon. These native short grasses require little water, no fertilizer, and only infrequent mowing.

What people do in areas close to their houses, however, is less of a concern to me than how they manage their larger property. In chapter 7 we consider overgrazing and the damage it does to the ecosystem and the health of the land. Overmowing, if I can make up a word, can be equally destructive. Native grasses, as discussed in chapter 7, do not tolerate being repeatedly cropped short during the growing season, whether by an animal or a mower blade. Doing so will eventually kill all but the shorter-growing grasses, leaving less desirable things such as annuals, sandburs, and weak, short grasses as well as more bare ground.

If native grass areas are to be mowed, it should be done as infrequently as possible (i.e., once a year, preferably when the grass is dormant in late December to early February). If a second mowing is done, the best time would be late June to early July, when most of the cool-season grasses are beginning to go dormant and before too much growth has occurred for warm-season grasses. Any mowing should be done at as high a setting as possible—at least four inches, and six inches is better. Unfortunately, most walk-behind mowers and riding mowers do not have a setting that high. It is especially important to have a high setting if there are wildflowers in the area, since most of them have green growth in the winter that will be set back if they are cut.

I have tried to discourage mowing native grasses, but everyone should also note the importance of not allowing grass or forbs to grow up adjacent to structures where a grass fire can catch the structure, as discussed in detail in chapter 16.

Agriculture and Wildlife Tax Valuations

Almost all agricultural land in the state is assessed for tax purposes at values reflective of its agricultural income-producing potential rather than the actual value of the land were it to be sold. The reason for this is that land values now exceed the amount indicated by the potential agricultural income. Were it not for the much lower valuation for tax purposes, property taxes would make farming and ranching in Texas impossible. (Note: One frequently hears the erroneous term "ag exemption." It is not an exemption; farmers and ranchers still have to pay taxes.)

Without going into all of the details, what is necessary to obtain or maintain an agriculture tax valuation is to be engaged in the production of an agricultural product (e.g., food, fiber, forage) and to do so at an intensity common to the area. It is not necessary to make an actual profit. Many small landowners wind up damaging their land under an ag valuation by overgrazing.

Several years ago a constitutional amendment was passed that allowed landowners who already had an ag valuation to switch to what is called a wildlife valuation. Under a wildlife valuation, landowners manage their land for the benefit of wildlife and are no longer required to have livestock (although they are still allowed to do so if they wish). They are required to prepare a wildlife management plan and to engage in certain activities beneficial to the wildlife species they are managing for. This system has allowed many small landowners to manage their land in a way that is better for the general health of the habitat, thus preventing it from becoming overgrazed or overbrowsed. The person responsible for administering these valuations in each county is the county chief appraiser. Many appraisers are very helpful to small landowners seeking to understand and comply with the requirements for these programs, although some do not view the wildlife or agriculture valuation favorably. Private lands biologists with the Texas Parks and Wildlife Department can also be helpful in developing land management plans for wildlife valuations, and there are even commercial businesses that prepare applications for landowners. In addition, Cibolo Nature Center in Boerne hosts workshops for landowners interesting in converting their land to wildlife valuation. For more information, see www.tpwd.state.tx.us/taxvaluation.

Conservation Easements

Many folks have spent years being good stewards of their land, which they dearly love. And, looking around at what is happening to other properties around them, being subdivided into smaller and smaller parcels, with some of the properties being abused or poorly managed, they become concerned about what will happen to their land after they are gone. About the only way most people can control what will happen to their property after they no longer own it is to place certain restrictions (such as the number of homes that can be built on it) on the property by donating a conservation easement on the property to a nonprofit land trust. These easements outline what the landowner does and does not want done on the property in the future and thus allows the current landowner to conserve the habitat in perpetuity, which the land trust then enforces. There can be tax advantages for the landowner to do this. It is not something everyone wants to do, but it is an option to be considered. In the interest of full disclosure, I should state that at the time of publication I am a member of the board of the Hill Country Land Trust.

22

How to Manage Your Property Better

When the last corner lot is covered with tenements we can always make a playground by tearing them down, but when the last antelope goes by the board, not all the playground associations in Christendom can do aught to replace the loss.

—Aldo Leopold, 1949

THROUGHOUT the earlier chapters, we talk about the ecology of the Hill Country, how it has changed since Europeans began settling here, and what problems we face as landowners in restoring or maintaining our land in a healthy condition, including overgrazing, overbrowsing, cedar encroachment, erosion, water catchment, oak wilt, exotics, and loss of native trees. We also discuss many things that we as landowners can do to improve the condition of our land.

Below is a list of eight general things you can do on your property to attempt to mitigate these problems. Following each is a list of specific actions you can consider taking on your property. Not all suggestions are appropriate for every property, and the priorities will be different for different people and different properties, but these are offered as ideas for you to consider, to talk to experts about, and to keep in mind as you live on and enjoy your beautiful Hill Country ranch.

1. Survey your property and its condition, critically, with respect to amount of erosion and loss of soil, grass cover and condition, amount of edible forage, amount of bare ground, condition and

Keeping a vision of what you would like your land to look like will help you make the decisions necessary to achieve that dream. These photos are from the Kerr Wildlife Management Area, where TPWD scientists can teach us much about land management in the Hill Country. Note the diversity of woody species, the good stand of native grasses, and the vegetation at all levels from the ground up to tree height.

diversity of trees and shrubs, number of surviving young trees, and amount of grazing or browsing relative to the property's carrying capacity.

- Walk your property; look for the amount of grass cover, height of grass, edible forbs, and bare ground. Take notes, take photos, run transects, make exclosures, and so on to record the current conditions, so that you can both judge the current conditions and note changes in the condition of your pasture with time.
- If you have animals, estimate the amount of actual grazable acres you have and divide that number by the number of animal units you have to get your stocking rate.
- Look for any root sprouts or saplings. Are they being eaten? If so, consider caging them. Make some exclosures here and there under hardwoods. If you have some small woody plants growing up inside agarita or prickly pear, be sure to continue to protect them.
- There should always be uneaten grass in the pasture, and not just threeawns, which nothing eats.
- After getting a good mental picture of what your place looks like, go out to Kerr Wildlife Management Area or other well-managed places and compare the condition of the landscape.
- Ask Texas Master Naturalists for a Land Management Assistance Program visit. See the State website—http://master naturalist.tamu.edu—and the Hill Country site, http://grovesite.com/tmn/hcmn

2. Institute management practices to control animal numbers to within the carrying capacity of your land, resting the land and/or rotating grazers if necessary. Never let the animals degrade the habitat.

- As a general rule, and there can be many exceptions, a pasture in moderately good condition in the center of the Hill

Country can have a carrying capacity of one animal unit for every twenty-five to thirty grazable acres. That means one mother cow with a calf will require twenty-five to thirty acres, a goat will require at least four acres (more if the population of white-tailed deer is large), and a horse around forty acres. Be careful not to count area under cedar bushes, steep rocky slopes, or land over a mile from water.

- Remember, when the stocking rate exceeds the carrying capacity, the range will be overgrazed and the carrying capacity will decrease even more. In the long run, the carrying capacity of the land will increase as the stocking rate is decreased.
- If your animal numbers are too high, then at least some of the animals need to be removed from the pasture part of the year, or penned and fed part of the time.
- Most exotics eat like goats, and large goats at that, so if you have exotics they have to be counted also.

3. Manage brush in a way that both provides food and shelter for wildlife and also maximizes the amount of rainfall that reaches the ground. In general, the ideal balance is to have cedar coverage of less than 20 percent on flat and gentle slopes, leaving most cedar on steep slopes to prevent erosion.

- Begin by surveying your property and deciding which individual cedar bushes, or which areas containing cedar, you want to clear. Then decide on the clearing method, weighing both the impact on the land and hardwoods as well as the cost.
- From a wildlife habitat standpoint, many animals do not like to be more than about 100 yards from cover/shelter, so leaving irregular patches of cedar in some places can be good.
- The main consideration when clearing brush is to "do no harm," which in most cases means mitigating any erosion. It is almost always best to leave cedar on steep, rocky slopes

because the erosion potential is so high there. In other areas, laying a single layer of cut cedar branches around on bare areas and across the slope slows any water movement and catches leaves and soil. The cut branches can also be piled in a circle to make an exclosure for hardwood sprouts or planted trees.

- If you have few hardwoods, trimming the bottom limbs of cedar to make a "tree shape" can be useful while still allowing grass to grow underneath.
- A common mistake people make is to remove all cedar, then ask what they can plant to block their view of the neighbors, the road, and other things they would rather not see.
- For a discussion of the habitat requirements of the golden-cheeked warbler, see chapter 18.

4. Follow all erosion control practices applicable to your property.

- Remember that the most important thing you have on your property is soil. Without soil you can't grow plants; without plants you can't raise animals. Also, as you lose more soil and plants, more rainfall runs off, and conditions just get worse and worse.
- The best thing you can have to prevent erosion is a good stand of native bunch grasses along with native trees, shrubs, and forbs. So the better your pasture condition, the less erosion. On steep slopes or along eroding creeks or draws, just about any living thing is better than nothing.
- Lay cedar limbs and posts across slopes, on bare soil areas, and in any small eroding areas. Low rock walls are also effective.
- If animal traffic is eroding a creek bank, fence them off and conduct water to a downslope trough.
- Survey your property annually, looking for evidence of mov-

ing water or eroding areas so you can address problems as they are just starting. Walk your property in a rainstorm, or just after, to see what is going on when lots of water is present.

5. Work to improve the grass quality and quantity in order to maintain the soil in good condition to best capture rainwater and to prevent erosion. A good stand of native medium to tall grasses is the best ground cover to do this.

- If you already have good grass cover, be sure to maintain it that way by not allowing any overgrazing or excessive mowing.
- If your property is overgrazed, or has been lately, then the best thing you can do for the land is rest it for a couple of years to allow native grasses to rejuvenate and make more seed. If you can't do this for the whole property, then try doing it for one section at a time.
- If you have bare areas where cedar once was, natural processes will probably fill in grass if the area is not grazed. Sowing small amounts of native grass seed may speed up this process.
- If you have bare areas where brush piles were burned, rake in adjacent healthy soil or mix in some compost or manure to speed up the reestablishment of soil microorganisms, and sow some native grass seeds around as well.

6. Be mindful of the possibility of oak wilt visiting your property, and minimize all activities that might create a new oak wilt center. Always paint all cuts and wounds immediately.

- Keep an eye on your trees and those of your neighbors for signs of oak wilt. The sooner you spot a new oak wilt center, the more effective any treatment you pursue will be.
- Be scrupulous about painting all cuts and wounds on all oaks in all seasons, immediately.

- If you have a red oak die that you suspect died of oak wilt, cut it down or girdle it to dry it out as soon as possible to prevent any fungal mat formation. Do not transport the wood to another location until it has cured for at least a year.
- If you have mostly live oaks with few other hardwoods, begin a program of planting a variety of native trees, so if the worst happens you will still have trees.
- If a valued live oak tree is threatened by an active oak wilt center moving to within 100 feet of the tree, know that timely treatment with Propiconazole 14.3 has an 80 percent chance of saving it.

7. Plant and cage native trees and shrubs in order to improve both species and age diversity and improve the general habitat on your property.

- The natural processes of replacement of older trees and shrubs with young ones have been disrupted by the overpopulation of deer and other browsers in the past thirty to forty years. The number of hardwoods as well as the number of species of trees and shrubs are declining, for the same reasons. In order to reverse this process on your property, plant a variety of native trees and shrubs, perhaps planting some every year for several years. In order to survive, most everything needs to be caged, at least until trees attain mature, tough bark and have most of their green leaves above the browse line. Getting native trees established is much more successful if they are planted in October or November and watered regularly for the first one to two years.
- Avoid planting any nonnative plants. Remember the laws of unintended consequences. Just because a tree or shrub or perennial is sold in a local nursery does not mean it is native or noninvasive.
- The more diverse the plants on your property, in both size

and species, the better wildlife habitat, the greater the diversity of butterflies, birds, and other animals, and the healthier your land will be.

8. If possible, work to manage the deer population at a level below one deer per eight acres in order to help return the habitat to its natural condition.

- The habitat around your place will not improve greatly unless the deer plus exotic population is controlled to below the carrying capacity, which is generally estimated to be one deer to between eight and twelve acres for this part of the Hill Country.
- Remember, each hunting license gives the hunter the right to take five deer, and you can have as many hunters on your place as you want. Unwanted, field-dressed carcasses can be donated to Hunters for the Hungry.
- There are no seasons and no requirements for taking exotics other than basic hunting safety regulations and a hunting license.
- It is difficult to manage deer populations on small acreages with low fences. It is possible, however, to form wildlife co-ops with your neighbors to make your efforts more productive.

Finally, good land stewardship requires continuing education. Fortunately, there are many opportunities to learn more from a variety of experts from different government agencies, such as Texas AgriLife Extension Service, Texas Parks and Wildlife Department, Texas Forest Service, Natural Resources Conservation Service, and numerous others.

23

Where to Go for Help

No one book and no one person can possibly answer all of the questions you may have about managing your property. But a variety of experts and other sources of information and help, all free, are available to you in the Hill Country. I urge you to take advantage of these people, whom you will find friendly, helpful, and knowledgeable. I firmly believe that learning how to manage a Hill Country ranch well requires continuing education, but it is also fun.

Government Agencies

The specialists working for these various government agencies are in general educated in biology or agriculture and have many years of experience, which makes them the most reliable experts in the area on most aspects of land management. Unfortunately, it is sometimes difficult to find the agencies in the phone book, because different phone companies list them differently. They could be in the regular alphabetical section, in the business section, under the State of Texas or the U.S. government. For that reason, in the lists below I include the state website, which can lead you to the local office, and if any of them change their website address you can always Google them. I listed them below the way my phone book lists them currently.

UNITED STATES GOVERNMENT OFFICES

Agriculture, Department of
Natural Resources Conservation Service
www.tx.nrcs.usda.gov
This is the former Soil Conservation Service, now called NRCS. The agents there are frequently the most knowledgeable about grazing issues, soil conservation, prescribed burning, and land management in general. They have offices in most counties.

STATE OFFICES

Texas AgriLife Extension Service (formerly Texas Cooperative Extension Service)
http://texasextension.tamu.edu
Many of us grew up calling these agents the "county agents." There is an office in every county, and they are most knowledgeable about plant diseases, how to grow crops and raise animals, which pesticides or herbicides to use for each problem, and so forth. They frequently organize seminars, classes, and field days that are open to the public and provide a lot of useful information. The Texas AgriLife Extension Service offices also sponsor the local 4-H clubs and the Master Gardener programs and cosponsor the Texas Master Naturalist program.

Texas Forest Service
http://txforestservice.tamu.edu/main/default.aspx
In the Hill Country, these agents spend most of their time on oak wilt issues and are therefore the most knowledgeable source about that and other diseases of native trees. This is a subject where there is a lot of misinformation going around, some provided by

people who claim to be professionals. I recommend listening only to those who have no economic interest in what you decide to do or not do. To find the office nearest you, go to the website and click on "Contact Us," then enter your location in the "Office Search" blank. The TFS has an excellent website devoted to oak wilt at www.texasoakwilt.org.

Texas, State of
State Parks and Wildlife Department
www.tpwd.state.tx.us
www.tpwd.state.tx.us/landowner

There are many divisions of TPWD, some devoted to managing our state parks and some as game wardens dedicated to enforcing game laws. The wildlife division has primary responsibility for managing and conserving the state's wildlife resources. TPWD biologists throughout the state provide assistance to landowners upon request. These are the people who are experts in habitat management, restoration and enhancement, protection of threatened and endangered species, and game management. Research at the Kerr Wildlife Management Area has yielded much of what we know about white-tailed deer in terms of food, nutrition, genetics, breeding, and habitat requirements in the Hill Country. This is, again, an area fraught with misinformation, so I urge you to listen to the experts. You can find the nearest TPWD office through the website. TPWD biologists can help with various land management issues, including preparing Wildlife Management Plans. The Kerr Wildlife Management Area gives tours and lectures that every landowner should see: call 830-238-4483.

Local Nature-Related Organizations

TEXAS MASTER NATURALISTS
http://masternaturalist.tamu.edu

As an original member and past president of the Hill Country chapter of the Texas Master Naturalists, I have to admit a certain bias in favor of this organization; it has provided me with a great deal of enjoyment, satisfaction, and education. The organization is sponsored by both the Texas AgriLife Extension Service and the Texas Parks and Wildlife Department and is patterned somewhat after the Master Gardener organization in that it requires attendance at forty-plus hours of initial instruction followed by advanced training and volunteer hours annually. The mission statement of the Texas Master Naturalists is "to develop a corps of well-informed volunteers to provide education, outreach, and service dedicated to the beneficial management of natural resources and natural areas within their communities." The goal of the initial instruction is to teach prospective members as much as possible about the local ecology and the various habitats that make up the local ecosystem as well as the plant and animal communities that inhabit it.

Some chapters (including the Hill Country chapter, which covers Bandera, Gillespie, Kendall, and Kerr counties) provide a service in which landowners can request volunteers to come to their property and help identify the plants growing there as well as discuss any applicable land management issues. If you are interested in such a visit or interested in becoming a member, you can find a contact for your local chapter by going to the state website and clicking on your county. Most chapters have monthly meetings that are open to the public.

Native Plant Society of Texas
www.npsot.org
"The purpose of the Native Plant Society of Texas is to promote the conservation, research and utilization of the native plants and

plant habitats of Texas, through education, outreach and example."
There are several NPSOT chapters in the Hill Country, and you
can get contact information on the one nearest you from the state
website. They all have interesting monthly programs on many
different topics related to native plants, nature, and ecology.

Nature Centers

Nature centers in the Hill Country provide collections of native
plants and native habitats from which visitors can learn, but they
also host numerous classes and presentations by experts on a
variety of nature topics. They also provide opportunities to learn
by doing volunteer work in a natural setting. Get on their mailing
lists to learn about upcoming events.

Cibolo Nature Center, Boerne, 830–249–4616, www.cibolo.org

Friends of the Fredericksburg Nature Center,
www.fredericksburgnaturecenter.org

Riverside Nature Center, Kerrville, 830–257–4837,
www.riversidenaturecenter.org

Lady Bird Johnson Wildflower Center, Austin, 512–292–4100,
www.wildflower.org

Bamberger Ranch Preserve, Johnson City,
www.bambergerranch.org, has become famous as a place where
an enlightened landowner transformed a dry, cedar-choked ranch
into a productive ranch with many springs and abundant wildlife.
Any of the tours and programs given there are well worth the
time.

Kerr Wildlife Management Area, 830–238–4483,
www.tpwd.state.tx.us/huntwild/hunt/wma, is a 5,000-acre research ranch in western Kerr County run by the Texas Parks and
Wildlife Department. Grazing methods, prescribed burns, and good
land management practices have been developed there, along with
considerable research on white-tailed deer. The facility gives tours
and seminars that explain its research findings and showcase one
of the best habitats in the area.

Enchanted Rock State Park in Gillespie County. Government agencies do a lot to help preserve the natural beauty of the Hill Country.

Native Plant Nurseries and Seed Sources

Finding a source of native plants and seeds is not always easy. Most nurseries have a few of the most common native trees and shrubs, but the natives are mixed in with many nonnatives and even some invasives. Furthermore, if asked for advice, some nursery personnel direct shoppers to whatever they need to sell the most or what they have the most experience with, regardless of its ultimate source. To make life even more complicated, some plant species grow in very large areas (little bluestem grows from here to Canada and points east), but although they are the same species, plants are not all the same ecotype, so a specimen native to North Dakota may not do as well in our hot dry summers as plants grown in this area.

Below I list the nurseries that I believe have the greatest number of mostly native plants of local origin. I do not list places that I believe are likely to have many exotic and invasive plants for sale and no priority on native plants. There will be exceptions everywhere, and I may well list places that do not in fact meet those criteria or fail to list some that do. Places change, my experience is certainly not perfect, and I apologize for a less than perfect list.

Above all, find out the scientific name of all plants you are considering buying and make sure that is what is on the label before you take it home.

Native American Seed, Junction, 800–728–4043, www.seedsource.com. You can rely on this company as a source of local seeds. I highly recommend that everyone obtain their catalogue, which has a lot of very good information on grasses and wildflowers, including how and when to plant.

Natives of Texas Nursery, between Kerrville and Medina on Route 16, 830–377–7683, www.nativesoftexas.com.

Medina Garden Nursery, 13417 St. Hwy 16, Medina, 830–589–2771.

Native by Native Landscapes, Johnson City, 866–868–9933, www.nativebynativelandscapes.com.

Websites. Identifying plants can sometimes be difficult. In addition to the books listed in the Bibliography, there are some excellent websites with good photographs of long lists of native plants. Here are a few of the better ones for Hill Country plants:

www.noble.org

http://plants.usda.gov

http://uvalde.tamu.edu/herbarium

www.wildflower.org/gallery

Appendix 1

Deer Browse Preferences

Adapted from "White-Tailed Deer Management in the Texas Hill Country," by W. E. Armstrong and E. L. Young, Kerr Wildlife Management Area, Texas Parks and Wildlife Department, September 2000.

Preferred Deer Browse

These browse plants usually show signs of being grazed even with moderate to low deer numbers. Presence of young plants of these species indicates low deer density and probably a good nutritional level.

Carolina buckthorn, *Rhamnus [now Frangula] caroliniana*
Cedar elm, *Ulmus crassifolia*
Chinaberry, *Melia azedarach*
Cockspur hawthorne, *Crataegus crusgalli*
Downy viburnum, *Viburnum rufidulum*
Littleleaf leadtree, *Leucaena retusa*
Slippery elm, *Ulmus rubra*
Texas kidneywood, *Eysenhardtia texana*
Texas madrone, *Arbutus texana*
Texas mulberry, *Morus microphylla*
Texas oak, *Quercus buckleyi*
Texas sophora, *Sophora affinis*
True mountain mahogany, *Cercocarpus montanus*
White honeysuckle, *Lonicera albiflora var. albiflora*
Wright pavonia, *Pavonia lasiopetala*

Good Deer Browse

Moderate to heavy grazing on these plants indicates moderate deer numbers. Numbers of these plants should increase with proper deer numbers.

Blackjack oak, *Quercus marilandica*
Carolina snailseed, *Cocculus carolinus*
Chinkapin oak, *Quercus muehlenbergii*
Common greenbriar, *Smilax rotundifolia*
Escarpment blackcherry, *Prunus serotina var. eximia*
Evergreen sumac, *Rhus virens*
Flameleaf sumac, *Rhus copallina*
Fourwing saltbush, *Atriplex canescens*
Heartleaf ampelopsis, *Ampelopsis cordata*
Ivy treebine, *Cissus incisa*
Lacey oak, *Quercus laceyi*
Mountain grape, *Vitis sp.*
Netleaf hackberry, *Celtis reticulata*
Poison ivy, *Rhus toxicodendron var. vulgaris*
Possumhaw, *Ilex decidua*
Post oak, *Quercus stellata var. stellata*
Roemer acacia, *Acacia roemeriana*
Saw greenbriar, *Smilax bona-nox*
Sevenleaf creeper, *Parthenocissus heptaphylla*
Skunkbush sumac, *Rhus aromatica*
Texas colubrina, *Colubrina texensis*
Texas redbud, *Cercis canadensis var. texensis*
Virginia creeper, *Parthenocissus quinquefolia*
White shin oak, *Quercus durandii var. breviloba*
Woollybucket bumelia, *Bumelia lanuginosa. var. oblongifolia*

Low-Quality Deer Browse

No moderate to heavy grazing of these plants should be observed. Moderate to heavy grazing indicates an overpopulated deer herd.

General condition of the deer herd will be poor.

Agarito, *Berberis trifoliolata*
Elbowbush, *Forestiera pubescens var. pubescens*
Fragrant mimosa, *Mimosa borealis*
Hercules club prickly ash, *Zanthoxylum clava-herculis var. fruticosum*
Live oak, *Quercus virginiana var. virginiana*
Netleaf forestiera, *Forestiera reticulata*
New Mexico forestiera, *Forestiera neomexicana*
Texas black walnut, *Juglans microcarpa*
Texas persimmon, *Diospyros texana*
Western soapberry, *Sapindus drummondii*

Little-Utilized Browse

Grazing on these species indicates extremely poor range conditions. Deer will be in poor condition with poor fawn crops, body condition, and antler development.

Ashe juniper, *Juniperus ashei*
Lindheimer's silktassel, *Garrya lindheimeri*
Lotebush, *Condalia obtusifolia*
Mexican buckeye, *Ungnadia speciosa*
Texas yucca, *Yuccarupicola*
Whitebrush, *Aloysia lycioides*
Willow baccharis, *Baccharis salicina*
Yucca, *Yucca sp.*

Note: Browse plants are placed in forage preference groups based on deer use of leafy material and not for mast preference. Deer readily eat acorns, persimmon fruits, mesquite beans, and cedar berries. Because of erratic rainfall patterns, mast is not always produced by the various browse species and is not considered a reliable food source for white-tailed deer. However, in many areas of the Edwards Plateau mast and fruit crops can become important food sources at critical times of the year. For instance, mesquite beans in the western plateau may be the primary food source during the winter period.

Common Forbs of the Edwards Plateau

PREFERRED FORBS

Arrowleaf sida, *Sida rhombifolia*
Blue curls, *Phacelia congesta*
Bur-clover, *Medicago hispida*
Dayflower, *Commelina erecta*
Engelmann's daisy, *Engelmannia pinnatifida*
Evening primrose, *Calylophus berlandieri*
Four o'clock, *Allionia spp.*
Indian mallow, *Abutilon incanum*
Knotweed leaf-flower, *Phyllanthus polygonoides*
Lamb's-quarter, *Chenopodium album*
Mat euphorbia, *Euphorbia serpens*
Maximilian sunflower, *Helianthus maximiliani*
Redseed plantain. *Plantago rhodosperma*
Spiderwort. *Tradescantia spp.*
Texas bluebell, *Eustoma grandiflorum*
Texas filaree, *Erodium texanum*
Trailing lespedeza, *Lespedeza procumbens*
Velvet bundleflower, *Desmanthus velutinus*
Wild lettuce, *Lactuca spp.*
Winecup, *Callirhoe digitata, C. involucrata*

LITTLE-UTILIZED FORBS

Basket flower, *Centaurea americana*
Blue flax, *Linum lewsij*
Bluebonnet, *Lupinus texensis*
Butterfly weed, *Asclepias tuberosa*
Clasping-leaf coneflower, *Dracopis amplexicaulis*
Columbine, *Aquilegia canadensis*
Coreopsis, *Coreopsis tinctoria*
Cowpen daisy, *Verbesina dentata*
Drummond's phlox, *Phlox drummondii*
Eryngo, *Eryngium*

Gayfeather, *Liatris mucronata*

Greenthread, *Thelesperma filifolium*
Horehound, *Marrubium vulgare*
Horsemint, *Monarda citriodora*
Huisache daisy, *Amblyolepis setigera*
Indian blanket, *Gaillardia pulchella*
Indian paintbrush, *Castilleja indivisa*
Lanceleaf coreopsis, *Coreopsis lanceolata*
Lindheimer senna, *Cassia lindheimeriana*
Plains bitterweed, *Hymenoxys scaposa*
Prairie larkspur, *Delphinium carolinianum*
Purple coneflower, *Echinacea purpurea*
Rain lily, *Cooperia drummondii*
Square-bud primrose, *Calvlophus drummondianus*
Standing cypress, *Loomopsis rubra*
Tahoka daisy, *Machaeranthera tanacetifolia*
Texas bluebell, *Eustoma grandiflorum*
Two-leaved senna, *Cassia roemeriana*
Yarrow, *Achillea millefolium*

Appendix 2

Common and Scientific Hill Country Plants Names

Grasses

Bermudagrass, *Cynodon dactylon*
Big bluestem, *Andropogon gerardii*
Blue grama, *Bouteloua gracilis*
Buffalograss, *Buchloe dactyloides*
Bushy bluestem, *Andropogon glomeratus*
Canada wildrye, *Elymus canadensis*
Cane bluestem, *Bothriochloa barbinodis* var. *barbinodis*
Common curly mesquite, *Hilaria belangeri*
Dallisgrass, *Paspalum dilatatum*
Eastern gamagrass, *Tripsacum dactyloides*
Green sprangletop, *Leptochloa dubia*
Hairy grama, *Bouteloua hirsuta*
Halls panicum, *Panicum hallii* var. *hallii*
Hairy tridens (Erioneuron), *Erioneuron pilosum*
Japanese brome, *Bromus japonicus*
Johnsongrass, *Sorghum halepense*
King Ranch (KR) bluestem, *Bothriochloa ischaemum* var. *songarica*
Kleingrass, *Panicum coloratum*
Knotroot bristlegrass, *Setaria geniculata*
Lindheimer muhly, *Muhlenbeergia lindheimeri*
Little bluestem, *Schizachyrium scoparium*
Meadow dropseed, *Sporobolus drummondii*
Plains lovegrass, *Eragrostis intermedia*
Purple threeawn, *Aristida purpurea*
Red grama, *Bouteloua trifida*
Rescuegrass, *Bromus unioloides*

Scribner dichanthelium, *Dichanthelium oligosanthes* var. *scribnerianum*
Sideoats grama, *Bouteloua certipendula*
Silver bluestem, *Bothriochloa laguroides* ssp. *torreyanna*
Southwestern bristlegrass, *Setaria scheelei*
Switchgrass, *Panicum virgatum*
Tall grama, *Bouteloua pectinata*
Texas grama, *Bouteloua rigidiseta*
Texas wintergrass, *Nassella leucotricha*
Tumble windmillgrass, *Chloris verticillata*
Virginia wildrye, *Elymus virginicus*
Willmann lovegrass, *Eragrostis superb*
Yellow indiangrass, *Sorghastrum nutans*

Woody Plants: Large Trees
American elm, *Ulmus americana*
American sycamore, *Platanus occidentalis*
Bald cypress, *Taxodium distichum*
Bigtooth maple, *Acer grandidentatum*
Blackjack oak, *Quercus marilandica*
Bur oak, *Quercus macrocarpa*
Carolina basswood, *Tilia americana* **var.** *caroliniana*
Cedar elm, *Ulmus crassifolia*
Chinkapin oak, *Quercis muhlenbergii*
Eastern cottonwood, *Populus deltoides*
Escarpment black cherry, *Prunus serotina* var. *exima*
Honey locust, *Gleditsia triacanthos*
Lacey oak, *Quercus laceyi*
Live oak, *Quercus fusiformis*
Mesquite, *Prosopis glandulosa*
Mexican white oak, *Quercus polymorpha*
Pecan, *Carya illinoinensis*
Post oak, *Quercus stellata*
Texas ash, *Fraxinus texensis*
Texas red oak/Spanish oak, *Quercus buckleyi*
Texas walnut, *Juglans microcarpa*

Woody Plants: Small Trees and Shrubs

Agarita, *Berberis trifoliolata*
American beautyberry, *Callicarpa americana*
Anacacho orchid tree, *Bauhinia congesta*
Ashe juniper, *Juniperus ashei*
Blanco crabapple, *Pyrus ioensis*
Buttonbush, *Cephalanthus occidentalis*
Carolina buckthorn, *Rhamnus caroliniana*
Cenizo, *Leucophyllum frutescens*
Creek plum, *Prunus rivularis*
Desert willow, *Chilopsis linearis*
Evergreen sumac, *Rhus virens*
Eve's necklace, *sophora affinis*
Flameleaf sumac, *Rhus lanceolata*
Fragrant sumac, *Rhus aromatica*
Goldenball leadtree, *Leucaena retusa*
Gum bumelia, *Bumelia lanuginosa*
Hop tree/Wafer ash, *Ptelea trifoliata*
Kidneywood, *Eysenhardtia texana*
Mexican buckeye, *Ungnadia speciosa*
Mexican plum, *Prunus mexicana*
Mexican silktassel, *Garrya ovata*
Pink mimosa, *Mimosa borealis*
Possumhaw, *Ilex decidua*
Red buckeye, *Aesculus pavia* var. *pavia*
Retama, *Parkinsonia aculeata*
Roughleaf dogwood, *Cornus drummondii*
Rusty blackhaw, *Virburnum rufidulum*
Spicebush, *Lindera benzoin*
Texas madrone, *Arbutus xalapensis*
Texas mountain laurel, *Sophora secundiflora*
Texas persimmon, *Diospyros texana*
Texas redbud, *Cercis canadensis* var. *texensis*
Toothache tree, *Zanthoxylum hirsutum*
White shin oak, *Quercus sinuata* var. *breviolba*
Yaupon, *Ilex vomitoria*
Yellow buckeye, *Aeschlus pavia* var. *flavescens*

Glossary

Animal Unit	Defined as a 1,000-lb cow with a calf up to six months old, which are assumed to eat 26 lbs of dry forage a day, or any number of other types of animals that eat 26 lbs of forage a day.
Annual	Plant that sprouts, matures, makes seed, and dies within one year. New growth comes only from seed.
Big Four	Four most prominent grasses of the "Tall Grass Prairie": big bluestem, little bluestem, yellow indiangrass, and switchgrass.
Biodiversity	Condition in which a given area has a large number of different plant and animal species. Biodiversity is generally considered the ideal, most healthy condition of a habitat.
Browse	Leaves of woody plants (trees and shrubs) that constitute the food for browsers, most notably white-tailed deer, most exotics, and goats.
Browse line	Notable line of demarcation about five feet high on woody vegetation, caused by browsing animals eating everything below that level.
Carrying capacity	Density of animals that a particular property or unit area can provide food, water, shelter, and cover for, long term, without the habitat being overused or degraded.
Cedar	As used in the Hill Country, refers to Ashe juniper (*Juniperus ashei*). This species is not a cedar but a juniper. A similar species, redberry juniper, grows in areas north and west of the Hill Country.

Cedar encroachment   Expansion of cedar, in both number and size of individual plants, which crowds out species that previously occupied the space.

Decreaser   Grass species that is most readily eaten by livestock, becomes reduced in size under over-grazing conditions and thus decreases in size and number. The term can also be applied to favorite woody plants and forbs. See **Increaser.**

Ecology   1. Branch of science concerned with the relationships between organisms and their environment. 2. Totality or pattern of relations between organisms and their environment.

Exclosure   Small fenced-in area used to protect plants inside from being eaten, or used to assess the amount of forage being grazed by comparing the amounts inside and outside the exclosure.

Exotic   Plant or animal not native to a general area, which is introduced, intentionally or accidentally, into the ecosystem.

Feral animal   Animal that was once a pet or livestock but has escaped human control and ranges freely, usually causing considerable habitat destruction. Feral hogs and feral cats are the worst offenders.

Forage   Food for grazing animals, usually grass or hay.

Forb   Broad-leaved herbaceous plant (not having woody stems). Most wildflowers and weeds are forbs.

Fragmentation   Larger ranchlands broken into many smaller parcels and sold to different individuals, resulting in a patchwork of land management styles that disrupt previous habitats for many species of wildlife.

Grazable acres   Number of acres on a given property that actually contain edible grass and are accessible to grazing animals, as opposed to the total acreage, which may contain areas not usable or accessible by the animals.

Habitat	Area in which animals live that provides them food, water, shelter, and cover.
High fence	Fence built high enough (usually seven feet or more) to prevent large ungulates, both exotics and white-tailed deer, from jumping over. Usually put up to keep exotic game animals in a given ranch. There is some controversy about high fences, because they disrupt normal wildlife movement.
Hill Country	That part of Central Texas within the Edwards Plateau bounded roughly on the east and south by Interstate 35 and the Balcones Escarpment from Austin to San Antonio and continuing West from San Antonio. On the north the boundries are less distinct but run roughly from Lampasas to Sonora to Rock Springs and Uvalde.
Increaser	Grass species that is less palatable or less nutritious but that increases in number and size in an overgrazed pasture. See **Decreaser.**
Invasive species	Any exotic species that reproduces and grows to such an abundance that it crowds out other species that would otherwise exist in the area. Although cedar and white-tailed deer crowd out other species, since they are native they are not normally referred to as invasive.
Monoculture	Solid stand of a single plant species. Not normally considered good habitat for wildlife.
Oak wilt	Disease principally of red oaks and live oaks, caused by a fungus that is spread by both small sap beetles and through the roots. Many live oaks in central Texas have been killed by this disease.
Overbrowsing	Condition in which browsers eat so much browse (leaves of woody plants) and forbs that the plants are not able to replenish what was eaten during the next growing season. This also results in small replacement shoots and saplings not growing into mature plants.

Overgrazing	Condition in which grass-eating animals eat so much grass that the grass plants are weakened and become less productive.
Perennial	Plant that has a normal lifetime of more than two years. Individuals die back seasonally but produce new growth from previous year's persisting part. All woody plants are perennials, but some forbs and many grasses are also.
Prescribed burning	Burning a pasture under certain, specified conditions of wind, relative humidity, and temperature, after thorough planning and preparation, by experienced individuals. The purpose is usually to improve the quality of the pasture.
Rotational grazing	Practice of moving herds of livestock from one pasture to another in order to rest pastures for as long as possible. This practice greatly increases the productivity and health of the land.
Ruminant	Grazing or browsing ungulate with a multicompartment stomach in which bacteria degrade the cellulose that is eaten so that the animal can absorb the resulting sugars. Cattle, sheep, goats, and deer are ruminants.
Stocking rate	Number of acres per animal or animal unit that are currently on a given field, pasture, or ranch.
Take half, leave half	Practice on well-managed ranches of grazing one half or less of the forage produced, which leaves the grass plants in a healthy condition to be able to grow even more forage the next growing season.
Ungulate	Hoofed herbivore such as deer, elk, antelope, sheep, goats, cattle, and bison.
Woody plant	Plant that is perennial and begins new growth on the previous year's stems; trees and shrubs.

Bibliography

General Topics

Bartlett, Richard C. *Saving the Best of Texas.* University of Texas Press, 1995. 221 pages.

Bedichek, Roy. *Adventures with a Texas Naturalist.* University of Texas Press, 1947. 327 pages.

Burton, L. DeVere. *Principles of Zoology and Ecology: Fish and Wildlife.* Delmar Publishing, 2003. 470 pages.

Butterfield, Jody, Sam Bingham, and Allan Savory. *Holistic Management Handbook: Healthy Land, Healthy Profits.* Island Press, 2006. 248 pages.

Holechek, Jerry L., Rex D. Pieper, and Carlton H. Herbel. *Range Management: Principles and Practices.* Prentice-Hall, 1995. 526 pages.

Leopold, Aldo. *A Sand County Almanac.* Ballantine Books, 1949. 295 pages.

Roemer, Ferdinand. *Roemer's Texas.* Translated by Oswald Mueller. Texian Press, 1967. 301 pages.

Schmidly, David J. *Texas Natural History: A Century of Change.* Texas Tech University Press, 2002. 534 pages.

Weniger, Dale. *The Explorer's Texas: The Lands and the Waters.* Eakin Press, 1984. 224 pages.

———. *The Explorer's Texas,* Vol. 2: *The Animals They Found.* Eakin Press, 1997. 199 pages.

Zim, Alexander C., Herbert S. Nelson, and Arnold L. Martin. *American Wildlife and Plants: A Guide to Wildlife Food Habits.* Dover Publications, 1961. 500 pages.

Animals

BIRDS

Rylander, Kent. *The Behavior of Texas Birds.* University of Texas Press, 2002. 431 pages.

Shackelford, Clifford E., Madge M. Lindsay, and C. Mark Klym. *Hummingbirds of Texas.* Texas A&M University Press, 2005. 110 pages.

Sibley, David A. *The Sibley Guide to Birds.* Alfred A. Knopf, 2000. 544 pages.

Tveten, John L. *The Birds of Texas.* Shearer Publishing, 1993. 384 pages.

INSECTS

Brock, Jim, and Kenn Kaufman. *Butterflies of North America.* Houghton Mifflin, 2003. 384 pages.

Drees, Bastiaan, and John Jackman. *A Field Guide to Common Texas Insects.* Gulf Publishing, 1998. 359 pages.

Glassberg, Jeffrey. *Butterflies through Binoculars: The West.* Oxford University Press, 2001. 374 pages.

Jackman, John. *A Field Guide to Spiders and Scorpions of Texas.* Gulf Publishing, 1997. 201 pages.

MAMMALS

Burt, William. *A Field Guide to the Mammals.* 3d ed. Peterson Field Guide Series. Houghton Mifflin, 1976. 289 pages.

Kays, Roland, and Don Wilson. *Mammals of North America.* Princeton University Press, 2002. 240 pages.

Schmidly, David J. *The Mammals of Texas.* University of Texas Press, 1994. 501 pages.

REPTILES

Conant, Roger, and Joseph T. Collins. *A Field Guide to Reptiles and Amphibians of Eastern and Central North America.* 3d ed. Peterson Field Guide Series. Houghton Mifflin, 1998. 616 pages.

Tennant, Alan. *A Field Guide to Texas Snakes.* Gulf Publishing, 1998. 291 pages.

Geology

Finsley, Charles. *A Field Guide to Fossils of Texas.* Gulf Publishing, 1989. 209 pages.

Lambert, David. *The Field Guide to Geology.* Checkmark Press, 1998. 256 pages.

Spearing, Darwin. *Roadside Geology of Texas.* Mountain Press, 1991, 418 pages.

Turner, Ellen Sue, and Thomas R. Hester. *A Field Guide to Stone Artifacts of Texas Indians.* Gulf Publishing, 1993. 393 pages.

Plants

FORBS

Ajilvsgi, Geyata. *Wildflowers of Texas.* Shearer Publishing, 1984. 414 pages.

Enquist, Marshall. *Wildflowers of the Texas Hill Country.* Lone Star Press, 1987. 275 pages.

Everitt, James H., Sale Lynn Drawe, and Robert I. Lonard. *Field Guide to the Broad-Leaved Herbaceous Plants of South Texas.* Texas Tech University Press, 1999. 277 pages.

Hart, Charles R., Tam Garland, A. Catherine Barr, Bruce B. Carpenter, and John C. Reagor. *Toxic Plants of Texas.* Texas Cooperative Extension, 2003. 241pages.

Loughmiller, Campbell, and Lynn Loughmiller. *Texas Wildflowers: A Field Guide.* University of Texas Press, 1999. 271pages.

Niehaus, Theodore F. *Southwestern and Texas Wildflowers.* Houghton Miffin, 1984. 449 pages.

GRASSES

Fort Hays State University. *Pasture and Range Plants.* 2d ed. Fort Hays State University Press, 2006. 176 pages.

Gould, Frank W. *Common Texas Grasses.* Texas A&M University Press, 1978. 267 pages.

———. *The Grasses of Texas.* Texas A&M University Press, 1975. 653 pages.

Hatch, Stephan L., and Jennifer Pluhar. *Texas Range Plants.* Texas A&M University Press, 1993. 326 pages.

Loflin, Brian, and Shirley Loflin. *Grasses of the Texas Hill Country.* Texas A&M University Press, 2006. 193 pages.

Rector, Barron S. *Know Your Grasses.* Texas Cooperative Extension, 2008. 100 pages.

WOODY PLANTS

Correll, D. S., and M. C. Johnston. *Manual of the Vascular Plants of Texas.* Texas Research Foundation, 1970. 1881 pages.

Cox, Paul, and Patty Leslie. *Texas Trees: A Friendly Guide.* Corona Press, 1988. 374 pages.

Diggs, George, Barney Lipscomb, and Robert O'Kennon. *Shinners & Mahler's Illustrated Flora of North Central Texas.* Botanical Research Institute of Texas, 1999. 1626 pages.

Simpson, Benny J. *A Field Guide to Texas Trees.* Lone Star Press, 1999. 372 pages.

Stahl, Carmine, and Ria McElvaney. *Trees of Texas.* Texas A&M University Press, 2003. 287 pages.

Tull, Delena, and George O. Miller. *A Field Guide to Wildflowers, Trees and Shrubs of Texas.* Gulf Publishing, 1991. 344 pages.

Vines, Robert A. *Trees of Central Texas.* University of Texas Press, 1984. 405 pages.

———. *Trees, Shrubs, and Woody Vines of the Southwest.* University of Texas Press, 1960, 1104 pages.

Wrede, Jan. *Trees, Shrubs, and Vines of the Texas Hill Country.* Texas A&M University Press, 2005. 246 pages.

Index

Notes